THE CLOCKWORK SPARROW

Time, Clocks,
and Calendars
in Biological Organisms

SUE BINKLEY
Temple University

PRENTICE HALL, Englewood Cliffs, New Jersey 07632

Library of Congress Cataloging-in-Publication Data

Binkley, Sue Ann.
   The clockwork sparrow : time, clocks, and calendars in biological
organisms / Sue Binkley.
     p. cm.
   Bibliography: p.
   Includes index.
   ISBN 0-13-073701-1
   1. Biological rhythms. I. Title.
QP84.6.B55 1990
574.1'882—dc19   89-3477
                                            CIP

Editorial/production supervision
  and interior design: Karen Bernhaut
Cover design: Ben Santora
Manufacturing buyer: Mary Ann Gloriande

 © 1990 by Prentice-Hall, Inc.
A division of Simon & Schuster
Englewood Cliffs, New Jersey 07632

The publisher offers discounts on this book when ordered
in bulk quantities. For more information, write:

     Special Sales/College Marketing
     Prentice-Hall, Inc.
     College Technical and Reference Division
     Englewood Cliffs, NJ 07632

Printed in the United States of America

10 9 8 7 6 5 4 3 2

ISBN 0-13-073701-1

Prentice-Hall International (UK) Limited, *London*
Prentice-Hall of Australia Pty. Limited, *Sydney*
Prentice-Hall Canada Inc., *Toronto*
Prentice-Hall Hispanoamericana, S.A., *Mexico*
Prentice-Hall of India Private Limited, *New Delhi*
Prentice-Hall of Japan, Inc., *Tokyo*
Simon & Schuster Asia Pte, Ltd., *Singapore*
Editora Prentice-Hall do Brasil, Ltda., *Rio de Janeiro*

To

Kathleen B. Reilly
and
Karen Mosher

CONTENTS

PREFACE

This book is about biological rhythms, principally the rhythms that are found to recur on daily or seasonal schedules. Most of the facts about these rhythms were established using animals, but humans have these rhythms too; they have been studied, and they influence our normal physiology, our behavior, and our health. Clocks and calendars pervade our daily lives. I will explain, however, that we, as other living things, possess an internal chronometer, a *biological watch*.

The purpose of this book is to provide an introduction that would allow almost anyone to bridge the gap from no knowledge of biological rhythms to an understanding that would permit them to comprehend a research article in the field. Thus, it should be useful to the general reader on the topic who wishes an understanding, and it should serve as an introduction to students who wish to become versed in the subject. The scope of the book is that the terminology of biology of time is described, the general rhythmic phenomena that exist are illustrated, and the underlying mechanisms are discussed.

The plan of the book is as follows: Chapter 1 introduces the subject, terms, and methods that have been used to obtain facts. Chapter 2 describes the key phenomenon; the fact that the rhythms persist in constant conditions, which is the essence of a *biological clock*. Chapters 3 and 4 explain how the rhythms are set by light, temperature, and social cues, and also how the rhythms are reset by time cues in the environment. Chapter 5 discusses unique circumstances, such as constant light, under which rhythms may be annihilated. Chapter 6 deals with the calendar—the biological progression of events such as hibernation that correlate with the change in seasons, the circannual clock that underlies seasonal cycles, and the daylength that controls seasonal rhythms. Not all cycles are circadian, and there are some aspects of timed behavior that have organization in addition to the daily time base. Chapter 7 deals with these—rhythms less than a day long that occur in sleep, lunar rhythms, tidal cycles, menstrual changes, and the temporal organization of behavior within a cycle. Chapter 8 deals with the underlying mechanisms: the clockshop of biological oscillators, the clockworks in the cells, and the hypothetical models, and the use of drugs to study how the clocks work. Biological organisms are distinguished from each other by their variation. Many of the phenomena in this book are general, that is, they occur in most organisms. However, it is useful to consider each species by itself to place its biological clock in the context provided by the framework of generalities. Chapter 9 surveys actual biological clocks discovered in some organisms—rats, hamsters, sparrows, chickens, roaches, mollusks, flies, and single-celled organisms. Chapter 10 focuses on humans, their biological watches and the potential and problems they pose. These potential applications for humans affect daily life, they have implications for shift work, they can be used when we travel east or west, and they will be important to consider for life in spaceships and on other worlds where the sun does not provide a 24-hour environment.

ACKNOWLEDGMENTS

I thank my colleague, Karen Mosher, for her work on the figures and proofreading. I thank my enthusiastic editors, Ken Tennity, Karen Bernhaut (editorial production), Karen Verde (copy editing), and Ben Santora (cover design). I thank the National Science Foundation which supported my research in circadian rhythms and the pineal gland for almost two decades. I especially thank Bruce Umminger, Elvira Doman, Charles Ralph, Shelley Binkley, and H. Randolph Tatem for personally encouraging my career. I thank Deborah Beaumont, Jocelyn McRae, and Kyra Tatem for library assistance and proofreading. I thank Dwight Crawford, Dr. Charles F. Ehret, Dr. Leland N. Edmunds, Dr. Franz Halberg, Dr. J. Woodland Hastings, Dr. Jon W. Jacklet, Dr. John D. Palmer, Dr. Shepherd K. Roberts, Dr. Beatrice M. Sweeney, Dr. Milton H. Stetson, Dr. Stephen Takats, Mária Begoña Tomé, Dr. Rütger A. Wever, and Dr. Arthur T. Winfree for the use of figures and for their suggestions on the scientific aspects of the text.

Sue Binkley
Temple University

PROLOGUE

This book is about time. It is about a very specific kind of timing that exists for plants, animals, and people. It is mainly about the timing of events that recur on a daily basis, but it is also about the biological events that recur on longer (seasonal cycles) or shorter (dream-sleep cycles) time bases. If you read this book, and have not thought about it before, you may arrive at a new appreciation of the role of time in your life and the possibilities for control of its use.

If you have thought about the subject of time, you realize that the word *time* has very broad usage. Webster lists no less than 29 definitions that occupy over a page of fine print in an unabridged dictionary. There are some very interesting kinds of time that this book is not about. For example, standing on the edge of the Grand Canyon listening to a park ranger attempt to explain the incredibly ancient ages of the layers of rocks cut by the Colorado River, you are faced with *geologic* concepts of time, the eons of time that constitute the history of the earth. I picked up a copy of Hawking's *A Brief History of Time* [176]. His discussion includes the *q* words of physics (quarks, quantums, and quasars) and cosmological concepts of time in which the direction of time depends on whether the universe is expanding or contracting. For Hawking, time is not an absolute as it was for both Aristotle and Newton, an unambiguous measure of the interval between two events.

You, the reader, may be relieved that this book does not require being able to visualize four dimensions of space and time. You have plenty of experience with the basic tool needed to understand this book, a clock on the wall or the watch on your wrist. You understand that *standard time* is an official time based on very accurate clocks kept by the government Bureau of Standards. This book is about the clocks that you, animals, and plants have in your bodies, the biological clocks. This book is not about the biological clock that runs out when you can no longer reproduce; it is not about the biorhythm charts that you can make with computers based, like astrology, on your time of birth.

In this book, I will show you many examples of organisms keeping time, in particular, being in possession of oscillations in their behavior and physiology. The reference standard for this time is the standard time, the clock on the wall, the watch on your wrist, that you are already comfortable using in your daily life. The time spans involved—seconds, minutes, hours, days, months, years—are all time spans with which you are familiar, and which, unlike geological time, take place within your lifetime.

You will immediately recognize that, except for Chapter 10, most examples in this book come from the plant and animal kingdoms. Like other areas of human medicine and psychiatry, our knowledge of how biological clocks work has its basis in observations and investigations with plants and animals. I am often asked how I became interested in this line of endeavor. The answer is probably that in the late 1960s I made the acquaintance of a drab colored but hardy, cheerful organism, the house sparrow, that could do remarkable things. First, it could be active 24h a day, month after month, with no apparent ill effects provided it was kept in constant light or pinealectomized. Second, without a watch or external time cues (such as dawn or dusk), using only that equipment with which it hatched, it was able to schedule its sleeping and waking with a near 24h period. My curiosity was so provoked by these phenomena that I have expended much of my waking time over two decades in pursuit of the mechanisms by which the sparrows accomplish their remarkable achievements. Throughout this book, I have used house sparrows for examples, as a *signature* organism for the study of the properties of biological clocks. You are not a house sparrow, but your biological clock has many of the same properties illustrated by the birds. This book then is about the *clocks* within you that wake you up before your alarm goes off, that give you times of *peak* ability during the day, that make you fall asleep at night. The same internal clocks make it difficult for you to adjust to travel across time zones or to shift work schedules.

1

WINDING UP
WITH
TERMS AND METHODS

"Where *rhythm* is a strictly descriptive word, *clock* is loaded with a functional connotation: it implies a device to measure time. And in this way it sets off a line of thought which *rhythm* fails to evoke. The line of thought concerns the functional prerequisites a clock must satisfy: it demands that the rate of the clock's time-measuring *motion* be independent of major environmental variables, like temperature, which are open to wide fluctuation; it demands that the clock be susceptible to synchronization with the cyclic external change to which it is functionally related."

C. S. Pittendrigh and Victor G. Bruce [262]

IS TIME THE SIXTH SENSE?

Biologists generally recognize five senses—hearing, smell, taste, touch, and vision. As noted by Keeton [195], sometimes the phrase *sixth sense* has been used to denote various kinds of extrasensory perception, but in fact humans and animals possess many senses. Perhaps one of these senses should be considered the *sense of time*. Even more appropriately, we might recognize senses of time because there are so many ways in which organisms use time in their lives. In the field of biological clocks, investigators devoted their efforts to discovering how plants and animals measure time, how they use time information, and how they organize the timing of their lifestyles and internal function. A consequence of this interest has also produced information regarding what happens when an organism's temporal order is disrupted.

Blatant clues to what kinds of timing might be important to organisms come from the dramatic and timed changes that recur in the earth's environment, such as the rising and setting of the sun, the ebb and flow of the tides, and the progression of the seasons. Artificial lighting, jet travel, and space exploration permit sudden disruptions of these natural temporal sequences.

The importance of clocks and calendars is obvious. Watches are probably the most common device carried by people, and we make daily use of calendars. The use of these items is so common in our daily lives that most of us take our use of time for granted. It is my opinion that timing is a factor that produces *stress* in modern cultures—the pressure to be on time and not to be late, the necessity to be in the right place on the right day, and the need to meet deadlines. (A deadline originally designated a line in a military prison which, if overstepped by a prisoner, would result in the prisoner being shot by a guard.) Our primitive ancestors may not have been under so much stress to keep track of time as we are; however, it would have been advantageous to them to have some sense of time of day and season of the year. Simple environmental cues were available to them. The position of the sun indicates time of day, and the length of daylight is a consequence of time of year. There is archeological evidence that our forebears went to considerable lengths to measure time. For example, Stonehenge, whose construction was a considerable undertaking, has been called a sun temple. Based on alignments of the stones, astronomers have accepted its purpose as a solar and lunar observatory.

However, the purpose of this book is not to discuss human history of time measurement, but rather, the remarkable internal chronometers possessed by most organisms. Using these biological clocks, most organisms can keep track of time, can organize their physiology in daily and seasonal rhythms, and can synchronize these rhythms with the prevailing external time by use of cues such as environmental lighting.

WHAT IS A BIOLOGICAL CLOCK?

Coleman [104] defined a biological clock as "an innate physiological system capable of measuring the passage of time in a living organism." His definition describes an *interval timer*, which simply measures the duration of a period of time, much as the sands run out in an hourglass. Presumably, his definition also includes the more rigorous concept of a *biological clock* that has been developed by investigators, especially in the field of circadian rhythms, which deals with cyclic events with a period of about 24h (as suggested in the quotation that opens this chapter). The characteristics the circadian investigators have listed for biological clocks go well beyond the simple ability to time an interval.

First, the central idea of a circadian biological clock is the notion that it is *endogenous*, that it is possessed by an organism. An organism is capable of generating a rhythm (recurring cycles), which has period lengths near 24h that are determined innately by some mechanism within the organism. Acceptance of this viewpoint was not a simple matter but has involved scientists in controversy. Proponents of the innate, or endogenous, biological clock have cited as their principle bit of evidence the fact that individuals generate rhythms with individually distinct period lengths. For example, one house sparrow might have a 23h rhythm, another house sparrow might have a 25h rhythm.

Second, a biological clock is nearly *temperature independent* or temperature compensated. This is quite surprising since many aspects of living organisms are dependent upon chemical reactions, and chemical reactions usually depend on temperature. Generally, a reaction proceeds faster at higher temperatures and slower at colder temperatures. From a functional viewpoint, temperature compensation may be a requirement for the biological clock, because the clock's utility would be reduced if weather changed the speed with which it ran.

Third, many investigators view the biological clock as a *continuously consulted chronometer*. That is, they think that an organism can use its circadian biological clock to tell what time of day or night it is, much as we consult a watch or clock to find out the time. An example that led to this idea comes from honeybees, which possess a time memory (*Zeitgedächtnis*); when bees find a source of nectar, they can return to it at the same time of day at which it was originally discovered.

Fourth, to be useful to an organism, it is reasonable that just as we must be able to reset our watches and clocks, so also must there be mechanism(s) for *resetting* biological clocks in order to synchronize them with events in the environment.

THE *WHY* QUESTION

"*Why* do we need a biological clock?"
"I don't know," whispered the professor.
"You don't know!" repeated the student.
"Not really," said the professor, a little more loudly.
"What am I supposed to put on the exam?" cried the student.

"Well, I wouldn't ask you for an answer to that question, I'd just expect you to be able to discuss it," the professor, pulling his beard, suggested with more confidence.

"Oh, you mean you want me to be able to list the evidence for the various theories, then?"

"Well, there isn't much real evidence, but we think a time sense is very important. *Why* do you think we need a time sense?"

"That's easy enough, to get to the exam on time. But I can use my watch for that," argued the student. (Fig. 1.1)

Figure 1.1 Evidence of the importance of time is that the item of equipment most frequently carried by human beings is a *mechanical watch* that tells the time of day. We *set* these mechanical watches to standard time and we *synchronize* our mechanical watches with one another. A flower doesn't need a mechanical watch; plants have their own internal time measuring devices or *biological watches*. In fact, so do animals and humans.

"True, but the plants and animals don't have watches. However, the idea to synchronize (your arrival) with something else happening in time (the beginning of the exam) seems to me to be something important for which a sense of time might be helpful."

"My distant ancestors didn't take exams. For that matter, animals don't wear watches, but they get along OK. Maybe they don't have a sense of time."

"We know they do. In fact, early evidence and most of what we know about biological timekeeping comes from scientists who studied plants and animals. We can speculate about why the biological clocks might have evolved. For example, take the seasonal clock; perhaps selection favored species that bred in the spring when there was plenty of food available for their offspring. Or, it might be of benefit to an insect to be able to time its visits to a particular species of flower to coincide with nectar production in that particular plant. Or, it might be helpful to be unavailable at the time your predator is on the prowl."

"This is starting to sound like a lecture." The student glanced at the clock on the wall.

The professor pulled a book off his shelf and continued, "Professor Bünning had something to say about why organisms use oscillations for chronometry. Here it is:

> 'In several cases, measuring time by oscillations certainly has advantages over using hour-glass processes. Oscillations allow *planning* not only for the next day, but for several days. However, in many cases of biochronometry, although using oscillations has no obvious advantage, it is not unusual to see organisms making use of them. [90]'

and . . . "

Seeing the professor was about to continue for hours, the student interrupted: "Is there any way to take away the biological time sense to find out what it does to organisms?"

"Well, yes, there are some ways to disrupt biological clocks."

"Well, well, you tell me, has anybody shown they are necessary for survival?" said the student, becoming belligerent.

"Maybe we're overlooking the obvious here," said the professor, edging toward the door.

"Wait, I want to ask you how the clocks work," begged the student.

"I don't know . . . " said the professor, consulting his watch, "but maybe we can talk about this another time."

EVOLUTION AND HISTORY

We can speculate that biological clocks themselves have an evolutionary origin. Since the rotation of the earth occurs on a 24h schedule, many environmental events recur each 24h: light, dark, warm temperatures, cool temperatures, etc. These conditions obviously affect organisms; for example, plants require light for photosynthesis. Thus, it is logical that living things have organized their physiology on time bases (daily, seasonal) that correspond to cycles present in the environment. It further seems reasonable that the ability to take advantage of the environmental cycles would be adaptive, and, in particular, that it would be useful for the organism to be able to predict the timing of the recurrence of environmental events since anticipation would permit preparation. Therefore, we can make an argument for why natural selection would favor the evolution of biological clocks capable of keeping track of and predicting the cycles that are present in the environment. Bünning discusses one hypothesis of how the actual selection of clocks might have occurred:

> "In order to develop a clock that allows an adaptation to the normal 24-hour periodicity, it was *only* necessary to select from the great variety of innate biochemical or biophysical oscillations. It was not necessary to *construct* special hour glasses. [90]"

His suggestion follows from the idea that one means to separate or compartmentalize incompatible events in a physiological process is to organize them with respect to time into sequences and oscillations.

Pittendrigh speculated on the order of evolution of functions for circadian oscillations:

> "There can be no doubt that the origin of circadian oscillations derives from the benefits of endogenous programming in relation to inevitable exogenous daily change in the life of cells living—typically—longer than 24 hr. . . . with its characteristic opportunism, selection has exploited the potential of a temperature-compensated circadian oscillator to perform other clock functions. Two of these, Zeitgedächtnis and sun-compass orientation, are limited to animals and clearly dependent on additional complexities of their central nervous system. Another, the photoperiodic time-measurement, is much more widespread but is clearly secondary . . . to the pacemaker's primary function of simple daily programming. [261]"

Daan and Aschoff [110] discuss the survival values of periodic functioning in organisms. First, they point out that the relationships of cycle lengths to body weight—lifespan, generation time, sleep cycle time, breath time, pulse time, etc., generally increase as animals have larger brain or body weights. Second, they note the advantage of synchronizing a timed event among members of a population. Third, they point out the advantage of temporal adjustments for energy expenditure, as in predation.

Measurement of time and prediction of timed events has early origins in the prehistory of man. Perhaps the reader has been interested in the mysteries of such monuments as Stonehenge, and in the hypotheses that some of these constructions were made and used to keep track of celestial events, seasons of the year, phases of the moon, and time of day.

Aschoff [14] traced the history of human interest in biological rhythms back to a Greek poet, Archilochus (who wrote about the rhythm governing man 2500 years ago); to a Frenchman, J. J. Virey (who described daily rhythms with the phrase *horloge vivante*, living clock, in 1814); and to D. C. W. Hufeland (to whom he attributes the 1779 suggestion that 24h is our natural time base). More direct discussions of the history of the field of circadian rhythms begin with the observations of the sleep movements of plants; leaves are spread out in the day to catch the sun, and droop at night (Fig. 1.2).

According to Bünning, whose *Die Physiologische Urh* was the first book in the field in 1958, the rhythmic movements were observed by Androsthenes during the march of Alexander the Great [89, 90]. However, reviewers [14, 89, 90, 368] also credit the discovery of persistent rhythms in leaf movements in sensitive plants in the dark to a French astronomer, Jean Jacques d'Ortous de Mairan, in 1729. The compendium of persistent rhythms found to exist in plants and animals accrues to this day. The examples used in this volume only

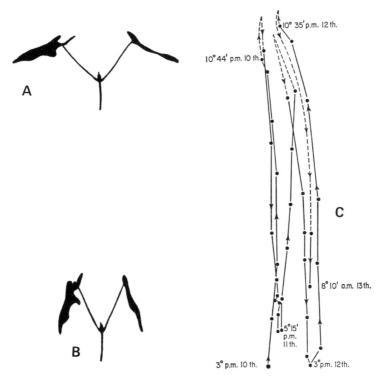

Figure 1.2 Rhythmic motions of plants are represented. The leaves of a bean seedling spread out to catch the sun in the day (A) and drop at night (B), which exposes less leaf surface. [87] Darwin recorded three cycles of movements of a tobacco leaf on July 10–13. (C) [115, 368]

begin to list the plethora of phenomena in the living world that exhibit persistent rhythms.

A history of the course of scientific inquiry would fill a volume in itself and it is not the purpose of this author to act as a historian. Much of the published work documents the existence of rhythms in various phenomena; the catalog of biological clock phenomena is ever growing. Yet other work has been directed to questions such as: How are the clocks set and reset? What are the environmental time cues and how are they perceived? What is the mechanism underlying the clock?

As the work developed, there was early realization that the circadian clock and daily events were important to an understanding of the way organisms perceive season. That is, there was not only a daily clock, but also a calendar that made use of the daily clock.

Last, when considering the impact of biological clocks on man, I suggest we must also consider a kind of *evolution* that has occurred in the way we use time in our lives. This evolution has been driven not only by our biology, but also by our technology (such as the use of artificial lighting and airplanes) and our occupations. To meet the temporal challenges of our current world we must be able to travel across time zones, we must be fit to do shift work, and we must adjust to work in space.

TERMINOLOGY AND ABBREVIATIONS

Folks who study biological rhythms have developed their own language, or jargon, to deal with the phenomena they observe and to permit discussion. The reader should not be discouraged by this; a small effort is required to understand the basics. The terminology becomes more understandable when actual data are examined than in its abstract explanations or diagrammatic representations. Here we attempt such explanation with apology and hope the reader will persist long enough to look at some data that show actual measured rhythms (Fig. 1.3).

Time can be thought of as a line extending backward (to the left) into the past and forward (to the right) into the future. We can represent a rhythm proceeding from left to right as a repeating waveform. A rhythm that persists in constant conditions is said to be *freerunning*. For example, if you descended into a cave and were thus isolated from daily light and temperature changes, and you removed your watch, you would have a rhythm of sleeping and waking that would continue, or freerun. The length of time between successive awakenings would be the *period* length of your freerun. The Greek letter tau (τ) is used by scientists to represent the length of a freerunning period. How often a cycle recurs, the number of cycles per day, is the reciprocal of tau, or frequency. In this book we will use period lengths to represent the lengths of rhythms.

When the timing of a freerunning rhythm is synchronized by a repeating external time signal, the rhythm is *entrained*. You are probably entrained in your normal daily life, probably by your use of an alarm clock. Plants and animals lack mechanical alarm clocks, but daily sunrise and sunset provide unambiguous time signals. The period length of the rhythm providing the signal is designated by the letter T. In our normal environment, T is exactly 24h. A signal that gives time cues has a special name, the German word for timegiver, *Zeitgeber*.

A particular time point of a rhythm is a *phase*. The time difference between two phases is called the *phase angle*. When a rhythm is altered so that its peaks occur later in time, the phase of a rhythm is *delayed;* when a rhythm is altered so that its peaks occur earlier in time, the rhythm is *advanced*. The application of these ideas is clear in people. When we travel east across time zones (as when an American flies to Europe), we encounter earlier sunrises with respect to our time back home. Our rhythms must *advance* in order to synchronize with the new time zone at our destination. When we journey

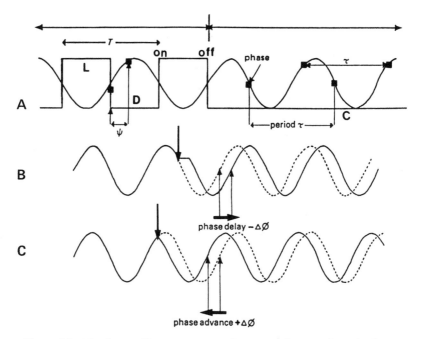

Figure 1.3 The figure illustrates terminology used for circadian rhythms. Here the biological rhythm is represented by a sine wave.

In (A), a square wave represents a light-dark cycle (L = light; D = dark) with a period length (T). In the normal world, T is determined by the rotation of the earth and is equal to 24 hours. The biological rhythm, represented by the sine wave, synchronizes with the light-dark cycle, or is *entrained* by the light-dark cycle (A, left). In constant conditions (A, right, small C), the biological rhythm proceeds from the time setting derived from the light-dark cycle and persists with its innate period (Greek letter tau, τ) near 24h. *Phase* (Greek letter phi, φ) is the location of a time point in the cycle; phase angle is the time difference between an arbitrary time point on the biological cycle and an arbitrary time point of the light-dark cycle.

(B) and (C) show changes in the timing of the biological rhythm (*phase shifts*) in which the biological clock is either set back (*delayed*, B) or set forward (*advanced*, C) by perturbations (vertical down-pointing arrows). (Redrawn after Saunders. [302])

westward across time zones (as when a European flies to America), a *delay* is required for our rhythms to synchronize.

CLEVER DEVICES

While the timing of human physiology is of most interest to the reader, the subjects for the majority of experiments with biological clocks have been plants and animals. Investigators have faced a particular challenge. In order to study a rhythm in some parameter, we are required to measure that parameter as time progresses. A collection of measurements taken at intervals as time progresses is called a *time series*. In order to obtain a time series, investigators have sometimes taken a sampling of a population of organisms at different times. Another method has been to continuously record a variable from an individual. Both methods have prompted scientists to innovate techniques. Although many of the studies have been done by round-the-clock sampling marathons by the investigators, automated sampling methods used in some studies have permitted the researcher to leave the laboratory. The best automatic methods are those that provide for the fewest breakdowns and minimum maintenance. In this book, you will see records that are months long and even some that were taken over years. The need for continuous monitoring has also predisposed the investigators to study those rhythms (e.g., locomotor activity, body temperature, Table 1.1) that can readily be monitored continuously over long periods of time without damage to the organism.

TABLE 1.1 Some Persistent Circadian Rhythms

Rhythm	Species
Petals raised in daytime	*Kalanchoe blossfeldiana* (plant)
Leaf drooping at night	*Canavalia ensiformis* (large beans)
	Phaseolus coccineus (beans)
	Nicotiana tabacum (tobacco)
Pale body color at night	*Ligia baudiniana* (isopods)
	Uca (fiddler crab)
	Anolis carolinensis (lizards)
Intense running activity at night	*Mesocricetus auratus* (hamster)
	Glaucomys volans (flying squirrel)
	Periplaneta americana (cockroach)
	Carcinus maenas (shore crab)
Morning emergence from pupae	*Drosophila* (fruit flies)
Night CO_2 fixation	*Bryophyllum fedtschenkoi* (succulent plant)
Bioluminescence at night	*Gonyaulax polyedra* (dinoflagellate)
Daytime perching activity	*Passer domesticus* (house sparrow)
Mating activity	*Paramecium aurelia* (protozoan)
Body temperature	*Myotis lucifigus* (bat)
Spore discharge	*Oedogonium cardiacum* (alga)
Phototaxis, peak in daytime	*Euglena gracilis* (infusorian)
Growth rate, peak in daytime	*Daucus carota* (carrot)
Volume of nuclei, peak in daytime	*Allium cepa* (onion)
Reduced night leaf heat resistance	*Kalanchoe blossfeldiana* (plant)
Night pineal melatonin synthesis	*Gallus domesticus* (chicken)
	Rattus norvegicus (laboratory rat)

An organism's movement from place to place, its *locomotor activity*, has often been the parameter of choice for study. We can use an animal's habits. For example, perching birds' locomotion can be easily studied in the laboratory by recording from electrical switches attached to perches in an individual's cage experimental timing regimens can be automatically supplied to the animals with timers (Fig. 1.4, 1.5, 1.6, 1.7).

Locomotor rhythms are readily measured in many species by attaching a recording switch to the wheel in a cage containing an individual or by making the cage a wheel. This method has been used to monitor the activity of rodents (rats, mice, hamsters) and even sea hares, lizards, and cockroaches! Activity can also be measured by placing an organism's cage on a knife edge so that its weight tilts the case against a switch each time the animal moves from one side to the other.

As mentioned (Fig. 1.2), plant leaf movements have rhythms that can be measured. This was done by fastening a string from a leaf to a needle that scratched smoke from paper on a revolving drum. Other rhythms, such as oxygen consumption or carbon dioxide release, can be measured by sampling the atmosphere of a plant subject. Yet another approach was used to measure the rhythms from populations of a dinoflagellate (single-celled alga) named *Gonyaulax*. *Gonyaulax* are marine organisms capable of bioluminescence— they glow at night. This is a rhythmic event that can be monitored in a culture of *Gonyaulax* by recording the light emanating from the organisms with photocells.

Figure 1.4 Photo of a detail of a watercolor of a male house sparrow painted by B. H. White when she was a graduate student.

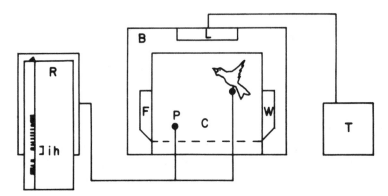

Figure 1.5 Diagrammatic representation of the apparatus used to monitor house sparrow circadian rhythms. Photographs of portions of the apparatus follow in this chapter.

An individual house sparrow was placed in a chamber (B) whose individual fluorescent lamp (L) was timed with a 24h timer (T). Food (F) and water (W) are provided ad libitum as is access to two perches (P) in the bird's cage (C). When the bird hops on and off its perches, microswitches open and close, breaking and completing an electrical circuit with a recorder (R). The recorder runs constantly so that a bird's record is obtained on paper as a series of pen marks.

Figure 1.6 Photograph of a Tork®timer. The *trippers* of the timer determine whether lights will be on or off in any 15 minute interval.

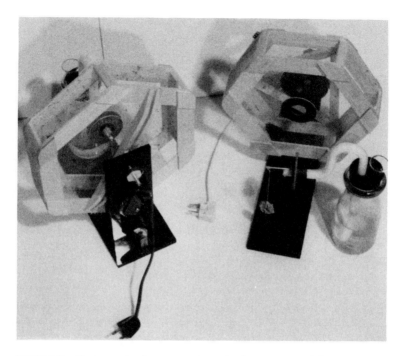

Figure 1.7 Photograph of two running wheels constructed for measuring circadian rhythms of live-in cockroaches. The wheel is made of lightweight balsa wood and mesh netting. To minimize friction, a sensitive magnetic reed switch detects rotations of the wheel. A pellet of rat chow is dangled in the wheel for the roach to eat and water is supplied via a cotton wick. (Courtesy of S. K. Roberts.)

Fruit flies have a rhythm of when they emerge, or hatch, from their pupal cases after metamorphosis, a process called *eclosion*. Flies emerging from populations of pupae stuck to tape can be gathered at time intervals by shaking them into collection vials for later counting. The name *bang boxes* was given one such apparatus so used.

We called our contribution to the catalog of rhythm-recording devices *pups in cups*, and it will serve as an example of the problems faced in studying rhythms. We were interested in recording the development of circadian rhythms in newborn laboratory rats (pups). The almost hairless, pink, two-inch pups are not able to run wheels when newborn, they cannot maintain body temperature, and they need to suckle for food. Our apparatus was simple and cheaply constructed (Fig. 1.8).

We made jiggle cages for individual pups by floating a styrofoam coffee cup (chosen for its light weight and low cost) on water in a glass beaker. The beaker was about one-half inch larger in diameter than the top of the cup so the cup leaned against the side of the beaker. When the pup was placed in the cup, its wriggling movements jiggled the cup lip against the beaker. To make

Figure 1.8 An apparatus designed to measure motion in rodent pups that were too young to run wheels. A styrofoam cup containing a pup was floated in a beaker of water. Aluminum foil was placed around the cup top and the beaker top so that as the pup jiggled the cup with its movements, an electrical circuit was broken. The pups require a warm environment, so the beakers sit atop a slide warming tray adjusted so that the temperature of the water in the beakers was 37°C.

electrical contacts, we lined the lips of the cups and beaker with aluminum foil. We ran wires to the foil on each lip so that as the cup jiggled it broke and made an electrical contact that we could record. Because the pups needed a supply of warmth, we placed the beakers on a histological slide warmer so the flotation water was maintained at body temperature, 37°C. The apparatus worked remarkably well, but we hadn't solved the food problem. Our preliminary solution was to rotate shifts of pups back and forth between the dams (rat mothers) and the cups. However, this procedure is undesirable experimentally because it means that the pups had an opportunity to derive time cues from their mothers. Nonetheless, we were able to record rhythms (or rather the lack of them) from developing rats using this technique (Fig 2.6).

Once a device has been invented that will monitor a rhythm, a recording device is needed to make a permanent record of the rhythm. Some rhythms have been recorded with an analogue tracing such as the bioluminescence rhythm. Event recorders, such as those made by Esterline-Angus® have commonly been employed by rhythm investigators to record simple on-off events (e.g., wheel revolutions, perch-hops, Fig. 1.9). The data so recorded have been displayed using raster (Figs. 1.10, 1.11, 1.12) or longitudinal plotting methods. Computerized technology (e.g., Mini-Mitter Systems and Dataquest® software, Sunriver, Oregon) are now being used by some investigators to accomplish the same task.

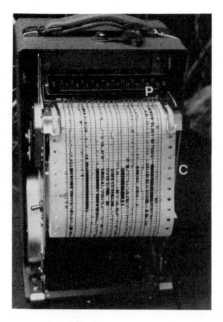

Figure 1.9 Photograph of an event recorder (Esterline Angus®) used to monitor locomotor activity from 20 animals, for example, from 20 house sparrows. Twenty pens (P) trace the individual perch-hops of the sparrows (horizontal pen strokes) on moving chart paper (C). Typically, the chart speed is 18 inches per day; a time marking (8 a.m.) appears left of the record for bird 1. The records are cut apart and dated at successive midnights.

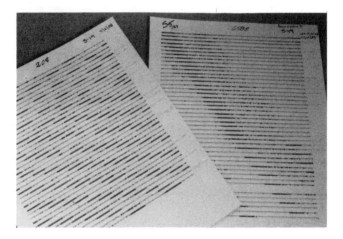

Figure 1.10 Photograph of a data card. The 20 bird records for a day (midnight to midnight) are cut apart using razor blades. Each bird's day of data is then pasted with rubber cement onto its own card (railroad board or poster board printed with blue guide lines that don't show up in photographs). Measurements are made from these records. For publication of figures such as appear in this book, completed records are composited and reduced in size photographically using a camera (Repromaster®) and rapid developing system (Copyproof®).

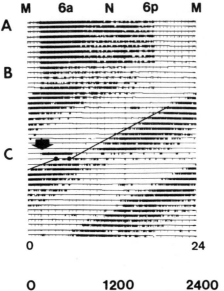

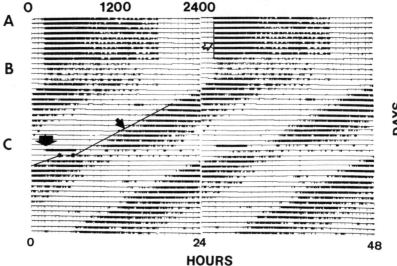

Figure 1.11 Locomotor record of a sparrow which illustrates the appearance of a finished activity record and which also demonstrates the phenomena described by the terms of Fig. 1.3. Horizontal labels above the upper record indicate Eastern Daylight Savings Time (M = midnight, N = noon). Each line of the upper record represents 24h; the lines are arranged vertically in chronological order (most recent line at the bottom). The lower record replicates the upper record. It has been duplicated horizontally so that the rhythm can be seen without interruption at midnight (where the records are cut). A line in the lower record thus represents 48h of data, and, as for the upper record, the lines are arranged vertically in chronological order. Time labels above the record

CAST OF CHARACTERS

Even being the anthropocentric organisms we are, the reader probably has already recognized the formidable problem of studying rhythms in the human. Who among us wants to be isolated, continuously monitored, and deprived of changes in environmental lighting? Fortunately, there have been some volunteers so that we have some data on human subjects (Chapter 10). But most of our knowledge of rhythms comes from plants and animals raised and drafted for the purpose of study by scientists. For most of the conscripted organisms, the experimental conditions provided compensations—a steady supply of food and water, a comfortable temperature, and protection from predators. An unnatural selection process favored the use of those organisms that were available and that thrived in the laboratory in long-term recording situations. Moreover, the organisms that displayed well-defined, easily measured rhythms were chosen. In addition, some organisms have been selected for a particular attribute, such as an accessible nervous system. All of these critters and many kinds of plants, along with humans, form the cast of characters for rhythm investigators (Table 1.1).

Laboratory rodents have been available to rhythm investigators; therefore, many studies have been done with laboratory rats and hamsters whose wheel-running activity can be monitored. Studies of these laboratory species of rodents have been augmented by investigations of many wild-caught species of rodents, for example, white-footed mice (*Peromyscus*). Indeed, some rat records, those of C.P. Richter, represent rats caught living in the wild in the environs of Baltimore. Most rodents studied have been night-active, or *nocturnal,* but there are some studies of small diurnal mammals, such as ground squirrels, and diurnal primates.

---

denote time on a 24h basis (1200 = N), (2400 = midnight). Time labels below the upper and lower records denote the number of hours in a line.

In A, the sparow was kept in a light-dark cycle (LD16 : 8) with lights-on at 2 a.m. and lights-off at 6 p.m.; the cycle was 24h long, thus T = 24h. The sparrow's activity synchronized (entrained) with the light-dark cycle—the sparrow was active in the light. The sparrow's perching began at 2 a.m., the time of lights-on, so the phase angle of the onset of activity (vertical line indicated by open arrow, an arbitrary phase reference point for the sparrow) with respect to lights-on (an arbitrary phase reference point for the light-dark cycle) was zero. At B the sparrow was placed in constant dark. Its perching activity continued to be rhythmic—the rhythm persisted (freeran) with a period length of 23h = τ. The rhythm drifts to the left because tau was shorter than 24h. Tau is calculated from a line drawn through the onsets of activity (filled small arrow). At C, the sparrow was perturbed with 4h of light (a single light pulse) beginning at 2 a.m. (filled large arrow). The sparrow was active throughout the 4h of light. Following the 4h light pulse, the sparrow continued to freerun in constant dark, but the phase of its activity onset was advanced by 2h (asterisks).

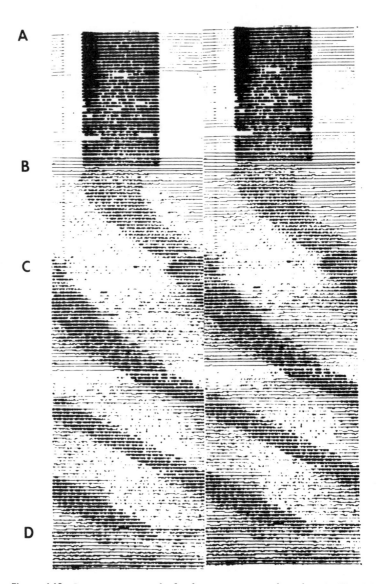

Figure 1.12 Locomotor record of a house sparrow plotted as in Fig. 1.11 (lower). In A, the sparrow entrained to LD12 : 12 (lights-on 5 a.m.; lights-off, 5 p.m.). At B, the sparrow was placed in DD. The DD freerun was longer than 24h (e.g., 24.8h) for most of the record. However, the tau in the first few days of DD was less than 24h, creating a *knee* in the record. At C, the sparrow was perturbed by an accidental light pulse. At D, the sparrow was pinealectomized, which disrupted its freerunning rhythm (discussed in Chapters 8 and 9).

Perching birds have been chosen because of their easily monitored activity and well-defined rhythms. The house sparrow, *Passer domesticus*, is considered to be a pest species and has worked well for us, but canaries, starlings, and other small birds have been regular subjects. As a group they are day-active, or *diurnal*. There is a dearth of studies of *nocturnal* birds such as nighthawks and owls.

Among the invertebrates, marine organisms (e.g., crabs) present special considerations because they may be subject not only to 24h fluctuations in light and temperature, but to 12.4h changes in the tides. In addition, many invertebrates have been attractive subjects because their physiology seems simpler and the components, especially of their nervous systems, are less complex than those of vertebrates. Marine mollusks, especially the sea hare (*Aplysia*) and its cousins, have gained attention for this reason.

Bean plants have been a favorite subject. However, the number of plant species that have been studied is legion. *Acetabularia* is a single-celled plant whose one cell has been of particular use because its cell nucleus resides in one region that can be removed. *Neurospora*, which is a bread mold, has been a favorite subject for geneticists. It has rhythms of sporulation that can be observed in a *race tube*. Growth medium is fixed to one side of a foot-long tube; *Neurospora* are added at one end; as the mold grows down the tube, it displays bands of sporulation at daily intervals. If it is grown in a dish, the bands of sporulation appear as rings (Fig. 1.13).

Lizards, reptiles, and fish, like the invertebrates, are subject to fluctuations in their body temperatures that are dependent upon the environmental temperature. Therefore, they are good organisms in which to study the role of temperature as a Zeitgeber. Many invertebrates and some vertebrates have vivid daily cycles in their body coloration because of melanophore cells in their skins. For example, a common lizard sold in pet stores is *Anolis carolinensis*,

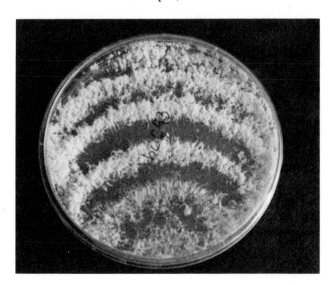

Figure 1.13 Photograph of a glass Petri dish in which *Neurospora*, a bread mold, exhibits rings which are formed at circadian intervals. The molds have been used to study the genetics and biochemistry of circadian rhythms. (The plate was a gift to the author from S. Brody.)

the so-called Florida chameleon. This lizard can rapidly change its skin color. The skin color has a rhythm (Fig. 1.14); the lizards are green at night (in the dark). In the day, however, the lizards' skin color depends upon the background on which they sit.

A behavior that has intrigued biologists for centuries has been the ability

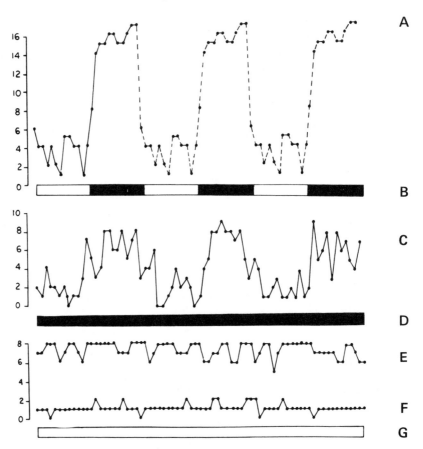

Figure 1.14 Rhythms of body color in lizards (*Anolis carolinensis*). The numbers on the ordinate represent the number of green lizards in a population. The horizontal axis represents 72h over which the colors of the lizards were observed at hourly intervals.

At the top, the lizards were kept in a light-dark cycle (LD12 : 12) indicated by the horizontal bar (B) beneath the graph of lizard color (A). The lizards were brown in the light-time and green in the dark-time.

In the middle, the lizards were kept in constant dark (bar D); a rhythm of color change (C, observed with a flashlight) persisted.

In the bottom, the lizards were placed in constant light (bar G). When the lizards were kept on a black background, they stayed green (E) and when they were kept on a white background, they remained brown (F). In either situation, they did not exhibit a rhythm of color change. (After Binkley, Reilly, Hermida, and Mosher [65].)

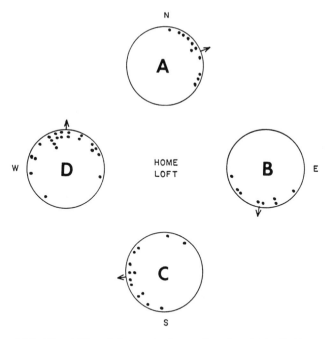

Figure 1.15 The ability of pigeons to choose directions is studied by examining the direction in which they disappear when released 30–60 km from their home loft (individual vanishing bearings indicated by the dots in the four circles representing four release locations—A, B, C, D; average vanishing bearings for each release point are indicated by the arrows). Normally, all the birds would fly in the direction of the home loft. However, the data here are for pigeons whose light-dark cycles had been advanced by 6h, shifting their circadian rhythms forward by 6h. The birds flew away in a direction that was 90 degrees in error from the direction of the home loft. The data show that the circadian clock is used by the pigeons in homing. (After Emlen [145].)

of birds to find their way on long migrations. Circadian investigators have proved that the circadian biological clock is used in direction finding (in species such as homing pigeons), so these organisms have contributed to our perception of the biological clock as a continuously consulted chronometer (Fig. 1.15).

Detailed consideration of the rhythms and biological clocks of rats, hamsters, sparrows, chickens, roaches, sea hares, *Gonyaulax*, *Euglena*, and fruit flies are the subject of Chapter 9.

CLOCKFACES

Scientists needed ways to represent and analyze biological time for circadian rhythms. This led them to develop various means of displaying data to look for rhythms. There are also statistical ways of examining data for oscillations. Since these methods permit us to look at a rhythm, I have placed them here under the heading clockfaces. LaMont C. Cole made a classic call for caution in analyzing biological rhythms in his paper "Biological Clock in the Unicorn". Using an example where he subjected random numbers to analysis and careful plotting, he pointed out the danger of the "possibilities for detecting *cycles* by means of relatively simple arithmetic procedures" [103]. His cautionary note is worth keeping in mind when faced with apparently arrhythmic data. Evaluation of a method means not only applying it to potentially rhythmic data, it is also useful to run data of known characteristics through any plotting or analytic procedure. For example, we can use artifically constructed time series of random numbers, as did Cole, or we can mathematically generate perfect rhythms—square-shaped waves, sinusoidal waves, and combinations thereof [30, 147, 148]. Such a procedure can be used to give us a context in which to evaluate biological data.

There is one aspect of the analysis of rhythms that is truly amazing. Simply stated, the amplitude of a rhythm is not as important as it might seem. For purposes of analyzing rhythms, it is often as sufficient to have the data in a binary form as it is to have a series of numbers showing the amplitude of the rhythm. This surprising idea, which seems to go against our intuition that the amplitude is important, has been discussed by J. H. Van Vleck [356]. The word *clipping* is used to describe the process of reducing data to binary form (zero crossings, events, on-offs, above-or-below an average). We tested Van Vleck's idea with sparrow perch-hopping data. When we analyzed sparrow data consisting of numbers of perch-hops (with the periodogram or autocorrelation described below), the results of the analyses were remarkably similar when we re-analyzed the same data after it had been clipped [31].

A question arises regarding how much data is needed for analysis. There is no hard and fast rule, and often investigators have been limited by constraints for measuring a particular rhythm or have been able to obtain very large quantities of data for a rhythm that is easily monitored. A rule of thumb that has been offered is that 10 or more cycles of data should be gathered [72]. A limitation occurs when we attempt to determine period length of freeruns. The limitation is that if the freerunning period varies over time, as it does when a sparrow makes a *knee* (Fig. 1.12), the assessment of period length is compromised.

There are two main methods for displaying circadian rhythm data. The first, the *longitudinal* method, is simply to plot measurements versus the time of day they were measured. The time of day is usually placed on the horizontal axis (Fig. 1.14). In the second method, the *raster* method, the horizontal axis is still time of day in the data array (Figs. 1.10, 1.11). However, the vertical axis is also a measure of time; 24h lines are arranged vertically in chronological order so that the vertical axis is time in days. The convention is to place the most recent day's data at the bottom of the array. In addition, sometimes the records are duplicated horizontally to permit a rhythm to be observed without interruption (Figs. 1.11, 1.12). The raster and longitudinal methods of plotting are sometimes combined (Figs. 1.16, 1.17). Both plotting methods also can be used to display rhythms with noncircadian periods (Chapter 7). Choice of the line length (e.g., 24h for circadian studies) for the raster method can magnify period changes.

An analytic method that is appropriate, not for detection of rhythmicity, but to describe many cycles of circadian data as one cycle, has been to make an *average curve*. To make an average curve, the period over which the average is to be made must be pre-selected by the investigator. Typical selections would be 24.0h for rhythms entrained to 24h regimens or the freerunning period (tau) for rhythms measured in constant light or dark. To make an average curve, an array of numbers is constructed as in the raster method, and the numbers are summed and averaged vertically for each time point (Figs. 1.18, 1.19). The averaging technique is useful for discovering the *shape* of the rhythm. As can be seen (Figs. 1.18, 1.19), an entrained daily rhythm is not sinusoidal; in the case

Figure 1.16 Body temperature record of a house sparrow. The sparrow's body temperature was recorded continuously using telemetry. To display the data so that circadian rhythms can be seen, the data were traced and shaded to an arbitrary baseline producing records that began and ended at midnight. Successive days of data were plotted one below the other with the most recent day's data at the bottom. In (A), the sparrow was in LD12 : 12 (cycle shown in bar over record); at (B), it was placed in constant dark and the temperature rhythm freeran with a period length greater than 24h; at (C) the sparrow was placed in LL and the temperature rhythm was attenuated [30].

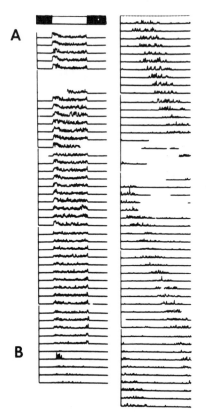

Figure 1.17 Activity of a house sparrow whose perch-hops were counted automatically at 6 min intervals. Each line of the record has 24h of data; the lines are arranged vertically in chronological order; the record continues in the second column on the right. In (A), the sparrow entrained to the LD12 : 12 cycle indicated by the bar over the record; note that the bird displayed more activity just after lights-on and just before lights-off than at mid-day, a *bimodal* pattern observed by the author in 56 percent of the sparrows kept in LD12 : 12. At (B), the sparrow was placed in constant dark and its rhythm freeran with a period longer than 24h. To make this record, the perch-hop 6 min counts were punched onto computer cards and plotted using a Calcomp® plotting system; gaps in the record resulted from technical problems in data collection. (After Binkley [30].)

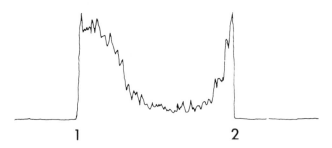

Figure 1.18 To make this graph, the activity of a sparrow whose perch-hops were counted at 6 min intervals was arrayed in 24h lines and summed vertically; thus the curve contains 240 points. The bird was kept in LD12 : 12, lights-on at (1), lights-off at (2). A point on this curve represents the average perch-hops that occurred at that time for 19 days. I called this an *average* curve. Note that it exhibits the post-lights-on and pre-lights-off increases in activity observed in 56 percent of house sparrows kept in LD12 : 12 (see Fig. 3.11). (After Binkley [30].)

shown, there are peaks of activity just after lights-on and just prior to lights-off (a *bimodal* pattern).

Seeking evermore objective means of assessing whether rhythms exist in biological data, Enright was the champion for a technique for analyzing rhythm data [147]. Basically, arrays or rasters of data are constructed and average curves made as described above. This is done for a selected series of period lengths (e.g., for Fig. 1.20, from 0.1 to 72.0h in 0.1h increments). The

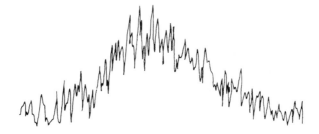

Figure 1.19   To make this graph, activity collected over 6 min intervals was averaged from a sparrow kept in constant light. The sparrow was freerunning so that its period length was 22.2h (not 24h); to make this curve, then, required making an array of the data in which each line was the length of the bird's circadian period length (tau). The curve for this sparrow, as for other freerunning sparrows, was a skewed sinusoidal wave; the bimodal pattern was absent. The graph spans 22.2h [30, 34].

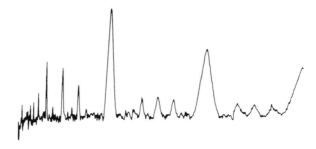

Figure 1.20   A periodogram for 16 days for a sparrow kept in DD. The periodogram is calculated by making a series of average curves of different period lengths (in this case, 720 average curves in which the data were arrayed in lines from 0.1 to 72.0 long), then plotting the variability that occurs in each average curve versus time. Thus the vertical axis (ordinate) for this graph is variation and the horizontal axis is time (72.0h). When an average curve has a lot of variation, as occurs when rhythmic peaks are aligned vertically, a peak appears in the periodogram. When an average curve has little variation, as occurs when the rhythmic peaks are not aligned vertically but *cancel* out, low values are obtained for the periodogram. Subpeaks may represent real rhythms, multiples and submultiples of rhythms present, artifacts, or a bimodal pattern. The tallest peak in the periodogram here represents the circadian rhythm present in the bird with a period of 23.9h [31].

amount of variation present in any one average curve can be calculated. When the selected period length equals the length of a rhythm in the data, the average curve will have the greatest variation. A plot of the variation versus the selected period lengths forms a graph, the *periodogram*. With this method, peaks occur in the periodogram at multiples and submultiples of a rhythm and due to the waveform so that interpretation of the subpeaks is not simple.

An *autocorrelation* method has also been used to analyze rhythms [72]. In this method, the data are arrayed in a longitudinal time series. As with the periodogram, a series of period lengths are selected (e.g., 0.1–72.0h in 0.1h increments, as in Fig 1.21). For each period length, every data point is multiplied by the data point that length away in time. If the points are maxima (peaks) of a rhythm, then the product is large; if the points are minima (nadirs), the product is small. For a given period length (called a *lag*), the products are summed and averaged. A plot of the resulting values is an *autocorrelogram*.

Various methods of *Fourier* analysis have been applied to obtain information about oscillations in data. I will describe the one used to obtain the *power spectra* in Figs. 1.22 and 1.23 [72]. To obtain these plots, sinusoidal-shaped waves were generated with formulas so that the period lengths of the waves ranged from 0.1h to 72.0h (in 0.1h increments). Each of these waves was multiplied against the previously calculated autocorrelation values. When the period of the formula-wave matched the period of the cycle autocorrelation, the product of the multiplications was larger than when the period of the formula-wave and the autocorrelation were not the same. A plot of the calculated values exhibits sharp peaks at the frequency (reciprocal of period length) of the rhythm; subpeaks appear when the wave-form of the autocorrelation function deviates from a sinusoidal wave.

The *cosinor* method consists of determining by least squares what cosine wave best matches a time series of data points. The method can be used with very few points (e.g., samples at 4h intervals) and only one cycle of data [170], but the meaning derived from the result is, of course, dependent to an extent on the number of time points available. Investigators using this technique have a conventional method for plotting their results on a circular graph representing 24h (Fig. 1.24). The advantage of the method is that it permits comparison of rhythms using a common set of specialized terms. The term *acrophase*, for example, means the phase of the peak of the best matching cosine wave.

Figure 1.21   Autocorrelation function of the same data record represented in Fig. 1.20. The graph is 72.0h long with 720 points. Peaks represent the circadian rhythm with a period length of 23.9h [31].

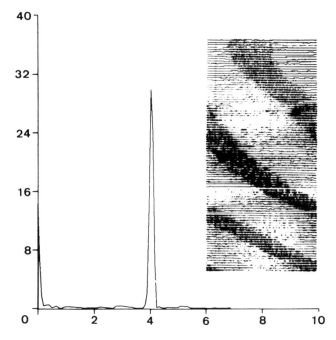

Figure 1.22 Power spectral analysis of 100 days of a sparrow's perching record in DD (using 6 min interval data collected from the same bird at the same time as the event record in Fig. 1.12). The large peak represents the freerunning circadian rhythm of the sparrow. The horizontal axis is frequency ($= 1/\text{tau}) \times 10^{-2}$, and the vertical axis is the power spectrum estimate $\times 10^2$ [30, 51].

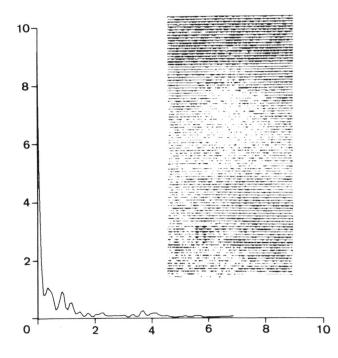

Figure 1.23 Power spectral analysis conducted on 100 days of a sparrow's perching in DD (data counted in 6 min intervals); the sparrow was pinealectomized and apparently arrhythmic. No large peak appears at the circadian frequency ($1/24 = 0.04$ cycles/h) supporting the loss of rhythmicity in the bird. The horizontal axis is frequency ($= 1/\text{tau}) \times 10^{-2}$ and the vertical axis is the power spectrum estimate $\times 10$ [30, 51].

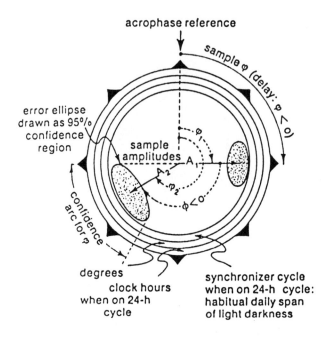

Figure 1.24   Representation of a cosinor analysis. Here, the circle represents 24h (like a 24h clockface) and the rhythms are represented by the arrows (A1,A2); variation in the data is represented by shaded error ellipses. (After Halberg [168].) In addition, the reader should consult an article entitled Chronobiology by Franz Halberg published in the *Annual Review of Physiology*, volume 31, 1969 which was selected as a 'Citation Classic.'

The compendium of methods for analyzing data is always incomplete. There are and will be new applications of old methods and application of new methods as new techniques are developed. I offer for an example of methodology with potential for analysis of biological rhythms the technology developed for *image analysis*. These methods can select out and measure particular *gray* levels (Fig. 1.25) and, in their more sophisticated forms, can improve the quality of images. The methods have been applied to improve photographs, especially in the field of astronomy, and to analyze the appearance of cells. For circadian rhythms, the methods promise a means by which the appearance of a rhythm in data may be sharpened, or improved. For this, as for other formalized methods of time series analysis discussed here, however, the reader is reminded that the bothersome *noise* present in the original biological data recording is part of the biological reality and is, in fact, what the organism is doing.

A   B

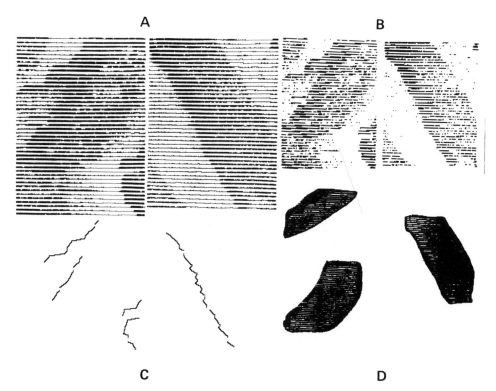

C   D

Figure 1.25 Representation of an image analysis of activity of sparrows (two-bird experiment discussed in Chapter 3). (A) shows portions of two locomotor records obtained from a pair of sparrows. (B) shows the computer image analysis of the records in which the computer has plotted all the data that are present at a *gray level* selected by the author. It is also possible to plot onsets (C) and to sum activity in selected regions (D). The system consisted of a video camera to depict the event record, an Apple® 2e computer, a HiPad® digitizing tablet, and Bioquant® software. (Binkley, 1988.)

2

THE
BIOLOGICAL
WATCH

"For years now I have felt that there is something wrong with our present concepts of time. A new concept of it could be one of the things that *is going to happen soon* . . . Fundamentally, I think I visualize it as some kind of absolute entity that moves inexorably past us; but on a newer and less secure level I am sure that it is more like a pattern through which everything we know about must move."

Colin Fletcher [157]
The Man Who Walked Through Time

FREERUN

The key observation that is the basis of the field of biological rhythms is that rhythms observed in nature continued under experimental *constant conditions* (unchanging temperature, lights always on, or lights always off). Scientists came to refer to this phenomenon as evidence for a biological clock. Some dictionary definitions of the word *clock* (see Glossary) differentiate a clock from a *watch*, which is a clock carried about on a person. Since most scientists believe that individual organisms are themselves able to generate rhythms, the biological clock should perhaps be thought of as a *biological watch*. The concept of an individual organism having its own biological watch is one that requires us to review our concept of the role of time in our lives.

A special word, *freerun* (which is not yet found in most dictionaries), has been coined to denote rhythms that are observed to persist in constant conditions. Examples of freeruns were already given in Chapter 1 (Figs. 1.11, 1.12, 1.14, 1.16, 1.17, 1.22); Fig. 2.1 illustrates two circadian rhythm freeruns mea-

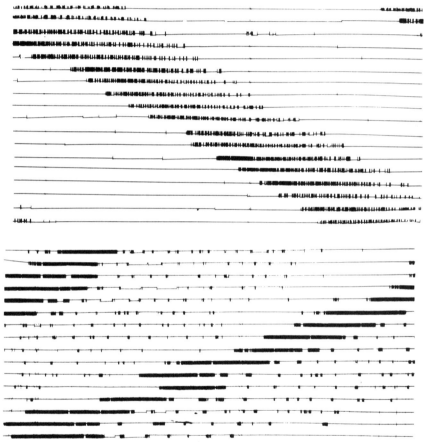

Figure 2.1   The upper event record shows a sparrow's freerun in constant dark for 18 days. The lower record shows a sparrow's freerun in constant light for 16 days. Note first that the period length of the freerun is longer in constant dark and that the period length is shorter in constant light. Note second that the duration of the activity is shorter in constant dark and longer in constant light. These features of sparrows' responses are characteristic of the freeruns of diurnal animals. Each line is 24h of data; the lines are arranged vertically in chronological order.

sured from two house sparrows. It is noteworthy that the period lengths of the freeruns are different from one another and, further, that the period lengths are not equal to 24.0h—the periodicity that would be present in the world excluded by the experimental system. In practice, investigators studying circadian rhythms suspect that a freerun whose period was exactly 24.0h would be deriving some 24.0h time cue from the environment and that the attempt to obtain constant conditions had not been successful.

Let us briefly review the terminology. The length of a cycle of a freerun is called its period (tau). The length of time in a cycle that the organism is active, its activity time, is designated by the Greek letter alpha (α); the length of time in a cycle that the organism is inactive, its rest time, is designated by the Greek letter rho (ρ).

Some animals can control the timing of events in the environment; for example, avoidance of the light-dark cycle might be achieved by a cave species. You are probably aware that, since people learned to use fire, humans have enjoyed mastery of the lighting and temperature in their environment. Our environmental time cues are potentially complex, being derived from the natural world and also our own artifice. We asked what an animal would do when allowed to control its lighting, and got more than one answer. Some animals left themselves in constant conditions, some animals did not appear to make use of the ability in any organized manner, but some animals chose a circadian rhythm of environmental lighting; that is, they turned on the lights during their daily active period (Fig. 2.2). In this chapter, as throughout this book, the emphasis is on circadian, or about-24h, rhythms. However, it is possible for rhythms that have time frames other than circadian—monthly or yearly—to freerun as well.

SPONTANEOUS CHANGES IN PERIOD LENGTH

If your watch was not accurate to 24.0h and if you wound it but did not reset it, your watch would freerun with a very accurate period. The biological watch is not as accurate as most mechanical watches in maintaining period length. When circadian rhythms have been measured over long periods of time, unexplained changes in period length have been observed (Fig. 1.12C). In sparrows, a maximum freerunning period occurred about 80 days after the birds were placed in DD, but the period lengths of mice freerunning in DD were still shortening when 300 days had elapsed [150]. The patterns of changes in freerunning period vary from one individual to the next. Explanations for spontaneous changes of freerunning period include changes in the environment, disturbances such as feeding, and aspects of development or aging. An intriguing possibility is the idea that the spontaneous changes reflect the biological oscillatory processes in the individual organism.

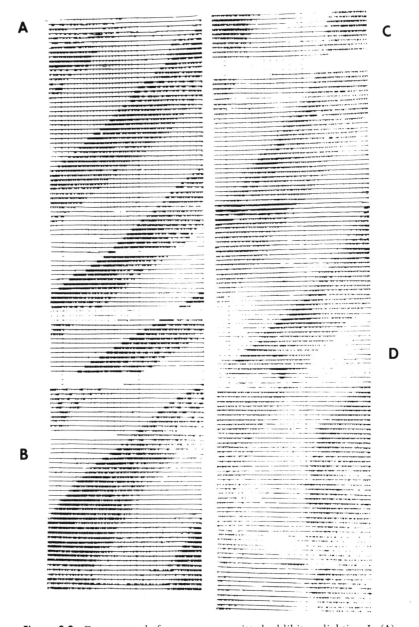

Figure 2.2 Event record of a sparrow permitted ad libitum lighting. In (A) and (C), the sparrow was given access to a third small perch that controlled the light in the sparrow's individual chamber (a hop on the perch turned on the light, the next hop turned it off) so that the sparrow could control its own lighting. The sparrow freeran. At (B) and (D), the light-switch-perch was disconnected, leaving the bird in DD in which its period lengthened. On the average, the sparrow's rhythms were 0.9h shorter when they controlled their own lighting [59].

AFTEREFFECTS OF PRETREATMENT CYCLES

Sometimes period length changes immediately following a change in environmental conditions—transfer from a light cycle into constant dark, or imposition of a light pulse (Fig. 1.12B, 1.12C). Collectively, observations of period length changes immediately following a pretreatment have been called *aftereffects* [256]. For example, sparrows often exhibit a period change when placed in DD from LD12:12 (Fig. 1.12); first the period shortens, then the period lengthens. This occurs frequently enough that investigators have given it a name; they say that a *knee* has occurred in the record. There are a number of observations that have loosely been classified among aftereffects.

First, aftereffects have been noted following imposition of a non-24h regimen (see Chapter 3). For example, period length in DD was larger after a 25h LD cycle than after a 23h LD cycle in hamsters. Second, aftereffects have ensued constant light pretreatment. For example, roaches and mice have a period in DD after LL pretreatment [245, 286]. Third, tau changes have been seen after light pulses (Fig. 1.12C has a second *knee*). Fourth, after 12–15h photoperiods, sparrows had shorter freeruns than after 2–6h photoperiods [150]. In a confirming study of sparrows, the period in DD was 23.7h after LD16:8 and the period length was 24.1h after LD8:16. Activity time was longer after LD16:8 than after LD8:16 [55]. Fifth, the sensitivity to light pulses (see Chapter 4) was altered by pretreatment photoperiod.

As with the spontaneous changes in period length, the most interesting speculations about the interpretations of aftereffects revolve about their possible reflection of underlying oscillatory processes, the biological clockworks (Chapter 8).

ACTIVITY TIME AND REST TIME

A freerunning circadian rhythm of locomotor activity can be divided into two parts—activity time (α) and rest time (ρ). It may surprise the reader that the durations of these times are subject to environmental control. Moreover, the effects of the environment are opposite for diurnal organisms (such as sparrows) and nocturnal organisms (such as hamsters).

In a light-dark cycle, a nocturnal animal is usually active in the dark. When the animals are placed in constant dark, the activity time is usually longer than it is for hamsters placed in constant light. On the other hand, the opposite is the case for a diurnal animal such as the sparrow (Fig. 2.1). Constant light increases the duration of activity (17h in Fig. 2.1) and reduces the duration of the rest period in freerunning sparrows as compared to DD (8h in Fig. 2.1).

The duration of activity in LL is a function of light intensity. For example, in an experiment with birds called chaffinches, *Fringilla coelebs L.*, the activity duration was 13.7h in dimmer light (1 lux) and the activity duration was 17.8h in brighter light (5 lux) [12]. Sometimes, it is useful to express the relationship of the duration of activity time to the rest time as a ratio, the alpha/rho (α/ρ) ratio. In the chaffinch experiment, the alpha/rho ratio in the dim light was +1.4, and in the bright light was −3.8. The ratios are usually negative for nocturnal species.

CIRCADIAN RULE FOR PERIOD LENGTH

Activity and rest durations are not the only parameters that are affected by the presence and intensity of light. Period length is also dramatically changed by the quality of lighting. Here again, nocturnal and diurnal species are usually affected in opposite ways.

For example, diurnal house sparrows usually have longer freeruns in constant dark and shorter freeruns in constant light (Fig. 2.1). Nocturnal hamsters have shorter freeruns in constant dark and longer freeruns in constant light (Fig. 2.3). Just as for the alpha/rho ratio, the intensity of the light is also important. The chaffinches in the same study mentioned previously [12]

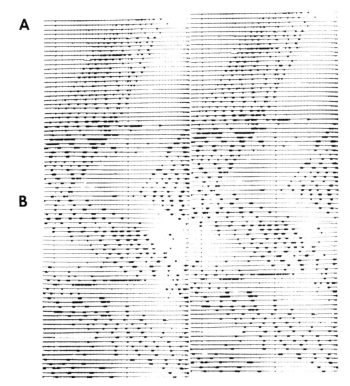

A

B

Figure 2.3 Event record of wheel running activity in a hamster (*Phodopus sungorus*). The hamster was kept in constant dark (A), where it freeran with a shorter period than it displayed in constant light (B). Note the immediate effect of the transfer to constant light. The effect, period lengthening in response to constant light, is typical for nocturnal species. Compare the data with diurnal response (Fig. 2.1). The record also shows activity organized into short bouts (Chapter 7). (Binkley, Mosher, and Spangler, unpublished data.)

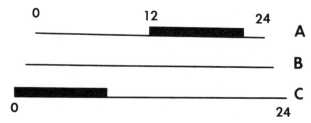

Figure 2.4 A diagram illustrating circadian time. Line B represents 24h of solar time. Line A represents circadian time for a 22h circadian biological clock such as that of a hamster in constant dark. By convention, c.t. 12 is set as the onset of wheel running in the nocturnal hamster. The c.t. time scale is divided into 24 parts, each part = $1/24 \times 22h$, or 55 min of the 24h time scale in B. Line C represents circadian time for a 26h circadian biological clock such as that of a sparrow in constant dark. By convention, c.t. 0 is set as the onset of perching activity in the diurnal sparrow. The c.t. time scale is divided into 24 parts, each part = $1/24 \times 26h$, or 65 min.

had a period of 23.4h in 1 lux LL and a shorter period of 22.5h in 5 lux LL. The modifying effect of light intensity upon period length of circadian rhythms has been called the *circadian rule*.

Circadian rhythm biologists have adopted a convention that is sometimes used to denote time of day on the biological watch (distinguished from time according to the sun or the clock on the wall). They have called this convention *circadian time* (Fig. 2.4). It is useful for designating time in freerunning rhythms. To determine circadian time, the scientist measures a freerunning rhythm. He then sets the period length of the rhythm (which is not 24.0h) equal to 24.0 units of time. By further convention, circadian time zero for diurnal animals is set at activity onset. Circadian time 12 is designated activity onset time for nocturnal animals. The application of circadian time can become confusing when making comparisons where alpha/rho ratios are different or photoperiods have been imposed, because the time of end of activity is not necessarily 12 for diurnal animals and the time of activity onset is not necessarily zero for nocturnal animals.

DEVELOPMENT

Scientists have asked when rhythms, especially circadian rhythms, appear during the development and maturation process. Human parents of newborns are only too well aware of the irregular and disturbing sleep patterns of their offspring during the initial weeks. When sleeping patterns of an infant were recorded for the first month of life, the circadian rhythm was not established at birth but was freerunning with a period about 25h by six weeks of age and gradually became shorter and more synchronized with a 24h schedule (Fig. 2.5) [117].

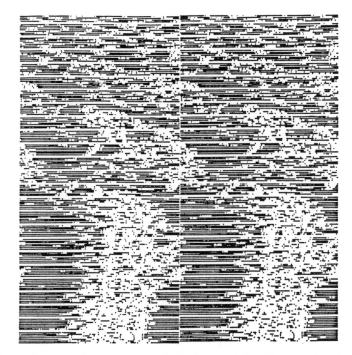

Figure 2.5 The sleep-wake record of a human baby. The record is double plotted so that the horizontal axis is 48h; the record extends from the second to the 26th week after birth. (After Kleitman and Engelmann [202].)

Freshly hatched chicks, *Gallus domesticus*, are precocious in that they can readily survive without a mother hen if kept in a warm environment and given access to chick feed and water. Daily cycles and circadian rhythms, for example of the pineal enzyme, N-acetyltransferase, can be measured in chicks as soon as they are exposed to a light-dark cycle. Indeed, when chicken eggs were kept in a light-dark cycle it was possible to detect a cycle in the pineals of the embryos as early as 17 days of incubation, well before hatching [45].

Offspring of many mammals are not as self-sufficient at birth as chicks. The pineal N-acetyltransferase enzyme rhythm appears at around four days of age in rat pups. We tried to measure locomotor rhythms of individual rat pups in constant conditions and isolated from their mothers; we found that movement rhythms were not present in the first 10 days of life (Fig. 2.6). However, rat pups have well-documented behavioral rhythms when they are housed with their mothers. This has led a number of investigators to study maternal–offspring interactions [117, 124, 182, 280, 332, 333]. Since rat dams will accept each others' offspring and their own offspring that were previously removed, it has been possible for scientists to conjure up ingenious experimental protocols based on transferring blind pups to foster mothers whose time settings have been changed (e.g., reversed with experimental lighting regi-

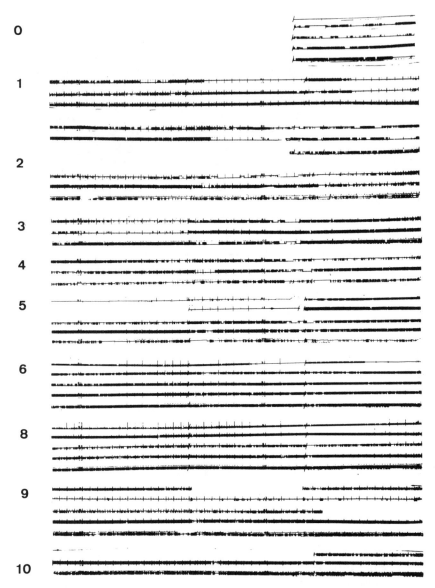

Figure 2.6 The early activity in rat pups measured using the pups-in-cups technique (Fig. 1.8). Each line is 24h of event data. The lines are grouped by the ages (0 to 10 days old, indicated on the left) of the pups from which the data were recorded. The pups did not exhibit any obvious circadian rhythm.

mens) [124, 332, 333, 334]. The thrust of these investigations is that nursing pups derive time cues (for the corticosterone, locomotor, water consumption, and pineal N-acetyltransferase rhythms) from the nursing dams with which they are housed, especially during the first week of life [301]. Feeding may be a possible cue for this since restricting the daily feeding to 4h of the day resets the locomotor rhythm [182]. Rat mothers have also been shown to synchronize the deoxyglucose uptake rhythm in the suprachiasmatic nuclei of the brains of fetal rats [280].

One means for maternal influence over rhythms of their fetuses may be hormonal, since it has been shown that hormones produced rhythmically by the dam (e.g., melatonin) are transmitted to the fetus. However, other possibilities exist, for instance, the locomotor activity or body temperature rhythms of the dams may act as rhythmic signals.

Thus, it would appear that the ability to generate measurable circadian rhythms arises at some point during development. Since rhythms occur in artificially incubated eggs, a parent is not required for the rhythms to develop. When a parent is present, such as a nursing rat, it can synchronize the rhythms of its offspring.

AGING

What is the effect of aging on circadian rhythmicity? The amplitude of many rhythms decreases as old age approaches. Usually this decrease is in a reduction of the maxima for the rhythm [278]. In addition to the occurrence of the amplitude-reduction phenomenon in hamster pineal N-actyltransferase (Fig. 2.7), the effect has been documented for other rhythms. A list collected from the literature by Davis [117] includes the rhythms in mouse body temperature, rat body temperature, mouse audiogenic convulsions, mouse oxygen consumption, human potassium excretion, human growth hormone, human testosterone, and human luteinizing hormone.

The amplitude is not the only property of a circadian rhythm that has been shown to change as age progresses. Pittendrigh and Daan [265] found that the freerunning period decreases in rodents from puberty to old age. The durations of activity and rest time change in humans over the course of their lives (Fig. 2.8).

There is evidence that disruption of 24h-entrained circadian rhythms

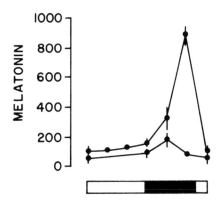

Figure 2.7 The life span of hamsters is a little over two years. The lower curve is melatonin in old (18 month old) hamsters. The amplitude of the melatonin rhythm was reduced with respect to the amplitude of the melatonin rhythm in young (2 month old) hamsters. (After Reiter and co-workers [278].)

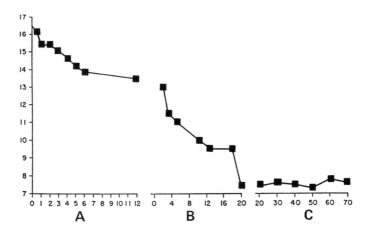

Figure 2.8 Hours of human sleep per 24h is plotted (vertically) versus age (horizontally) for infants (0–12 months, A), young people (0–20 years, B), and adults (20–70 years, C). Sleep is reduced markedly to between 7–8h as aging occurs. (After data of Webb [369].)

may be detrimental. Fruit flies subjected to non-24h regimens or constant light, or that are phase-shifted every two weeks, do not live as long [117, 269].

ACCURACY

We know how accurate our own mechanical watches and clocks are, and we *reset* them as needed to match standards of time provided by the government. Enright [148] considered the precision of the circadian biological clock and reported that three canaries, for instance, had individual cycle-to-cycle accuracies with standard deviations of three, four, and seven minutes. He described this as an error of as much as one part in 200 or as little as one part in 500. We looked at cycle-to-cycle variation in a sparrow and found that the difference was as small as 10 min or as large as 85 min (Table 2.1). DeCoursey's [120] flying squirrels are well known for their accuracy; they were more precise than our sparrows with individual variation as small as two min to as large as 15 min. However, the true precision of the underlying circadian pacemakers in the individual canaries and sparrows may be greater than the precision in the measured rhythm. This is because looking at the observable rhythm from which we can make measurements is like looking at the *hands* of a clock, not its underlying mechanism. In other words, the variation in the data is probably at least partly due to variability in the *indicator process* (here, perching activity) and not necessarily the *controlling process*. Enright [148] compared the precision of circadian rhythms to other rhythms—the spontaneous firing of single neurons (more variable than one part in 50), the human heartbeat (more

TABLE 2.1 Circadian Periods for Ten Successive Cycles in a Sparrow[1]

Cycle	Tau (h)	Difference (min)
1	24.50	30
2	23.84	− 10
3	25.02	61
4	24.32	19
5	24.65	39
6	24.43	26
7	24.41	24
8	24.96	57
9	24.27	16
10	25.42	85
Mean	24.68	37
Standard deviation	0.70	24
SEM	0.22	8

[1]The period for each single cycle (activity onset to activity onset in DD) was measured from the event record of a sparrow with clear perching activity onsets. The cycle-to-cycle differences from 24.0h in minutes give a measure of accuracy. There was a 95 min difference between the longest and shortest period lengths. Measurements were made from event records using a digitizing tablet and Bioquant software. (Binkley, 1988.)

variable than one part in 60), the chirping of crickets (one part in 30), flashing of some fireflies (one part in 200), and the discharge pattern of electric fish (one part in 8000). The entrainment process improves the accuracy, at least in the timing of the onsets of sparrows (to the extent that day-to-day variation in activity is less than we can measure from our recordings, or zero).

UBIQUITY

Pittendrigh lists empirical generalizations about circadian rhythms which includes the observation that:

"Circadian rhythms are ubiquitous in living systems . . . This holds in the systematic sense of kinds of organisms, and the physiological sense of kinds of functions. The emphasis in the literature on rhythms of say, locomotion and leaf movement reflects only ease of assay for these *superficial* phenomena; rhythms of DNA synthesis, e.g., exist but are less easily followed routinely." [256]

For the most part, where circadian rhythms have been sought, they have been found. Thus, there is a long list of organisms in which circadian rhythms have been studied. (Table 1.1; I counted over 150 species in one book [256] and almost a hundred in another [158].) There are also long lists of functions within individuals of a species that exhibit circadian rhythms [105, 215]. (Fig. 10.1.) In looking at these lists and the reports that generated them, we remember that in some cases the daily recurrence of an event has been shown but that demonstration of persistence of a rhythm has not been attempted.

TWO VIEWS

Early in the development of the field of circadian rhythms, two hypotheses for the mechanism responsible for the freerunning rhythms were put forth. Brown, Hastings, and Palmer debated the arguments for the two viewpoints in a book devoted solely to that purpose [85].

The first hypothesis, the *exogenous* hypothesis, was championed by Brown. Brown was impressed by some observations of crabs. Fiddler crabs, *Uca pugnax*, are dark colored in the day but turn pale at night. Captive crabs kept in the laboratory had two periods of locomotor activity each day about 12.4h apart that followed the timing of the tides on the crabs' home beach even when the crabs were kept in a light-dark cycle. Thus, it appeared the crabs were able to synchronize with some parameter in the environment that had the same frequencies as the tides. An interpretation of freerun, then, is that it originates from periodic events in the external environment to which the organism is synchronizing. Collectively, these potential time cues were called *subtle geophysical factors*. The subtle geophysical factors include such things as gravity, geomagnetism, atmospheric tides, electrostatic fields, and background radiation. Rhythms whose freerunning period was 24.8h were particularly suspect because that is the period imposed by the lunar day. The hypothesis of environmental timing of the freerunning clock is today the agnostic viewpoint, though the ability of organisms to detect subtle variations in their environment is not disputed. The proponents of the exogenous view point to the considerable body of evidence that organisms can detect such things as the local geomagnetic field.

Most circadian rhythm investigations have proceeded with their studies from an acceptance of the *endogenous* hypothesis [80]. In this hypothesis, individual organisms are viewed as in possession of an internal cellular-biochemical oscillator that is responsible for freerunning circadian rhythms. As Hastings describes, a piece of evidence that the endogenous supporters found compelling was that individual organisms freeran with period lengths that differed slightly from one another. The view of the scientists that support the *endogenous* clock is that environmental cycles are used as time cues that can synchronize, or reset, the internal clocks.

AN INHERITED CLOCK

Although there are some aftereffects of timed pretreatments, new frequencies do not seem to be learned. Subjecting plants or rats to 16h cycles did not cause them to lose their ability to produce a 24h rhythm [90]. In a recent study, mice were compared that were born and raised in a 20h or 28h cycle (instead of the normal 24h cycle) [116]. The freerunning periods, activity times, and rest times of these offspring mice were not affected by the exotic length of the cycles.

Raising successive generations of mice in constant light did not result in a loss of their abilities to generate circadian freeruns [90]. Second generation mice in one experiment had average taus of 25.5h, 26.1h, and 25.2h. Even keeping rats in LL for 25 generations did not cause them to lose their rhythms. The evidence for a hereditary circadian clock comes from many species—rats, mice, fruit flies, nightmoths, bees, chickens, lizards.

Another approach has been to select specimens with different periods and to study hybrids. With some species, intermediate periods were obtained, which leads to the conclusion that many genes participate. However, Konopka and Benzer made mutants of fruit flies with a chemical (ethyl methane sulfonate) [204]. The mutant flies had altered X-chromosomes and some were arrhythmic, some had 19h periods, and some had 28h periods. The investigators attributed their results to a single gene on the X-chromosome. As Konopka reviews [203], an amazingly similar locus [155] was also implicated for circadian rhythms of bread mold:

> "A genetic locus with certain properties similar to that of the *per* locus as been described in *Neurospora* . . . This locus can also be mutated to short and long period lengths . . . . there may exist loci with similar functions in both *Drosophila* and *Neurospora*; it is possible that the circadian pacemakers in these organisms are constructed similarly on the molecular level. [203]"

About 12 *Neurospora* clock mutants were found with either short or long taus; of these mutants, seven mapped to a single locus named *frq* (for frequency). Curiously, the mutants differed from the wild type by some multiple of 2.5h [154]. The studies of the genetics of circadian rhythms of fruit flies and bread mold echoed the 1972 conclusions of Bruce [86], who used another organism, *Chlamydomonas reinhardi*, and also championed a one-gene hypothesis for the period length.

3

SYNCHRONIZATION
WITH
THE ENVIRONMENT

"As in physical oscillations, the inverse of the period is the frequency, the number of cycles which take place in a given time. In the discussion of biological rhythms, frequency is used less often than period, however. Just as in physical oscillations, the frequency of rhythms may be forced to match exactly that of some external oscillation, the alternation of light and darkness, for example. When this occurs, the rhythm is said to be entrained by this external oscillation. The signal responsible for entrainment has been called the *Zeitgeber* or *time-giver.* Entrainment, of course, can only take place while the external oscillation is actually present. When a rhythm is not entrained, it is said to be freerunning, and now shows its natural period."

B. M. Sweeney [320]

DIRECT EFFECTS AND ENTRAINMENT

When some people first discover data showing a freerunning biological rhythm (see Chapter 2), they jump to the conclusion that the freerun represents the natural situation, and further aspire themselves to attain that state as somehow more natural and worthy. However, in nature, biological rhythms rarely freerun; they are usually synchronized by some external cycle that recurs in the environment (Fig. 3.1). In considering freerun and entrainment, I suggest that we keep in mind the possibility that the ability to freerun with a period (τ) slightly different than the period of the Zeitgeber (T) may be a prerequisite for the ability to entrain, an essential feature of the *clock*.

There are also responses to cycles that do not involve synchronization of an innate biological clock. I have called the nonclock phenomena *direct* responses [53]. An example that is important here is that some sparrows hop on their perches when the lights are turned on, and they stop hopping on their perches when the lights are turned off (Fig. 3.2). Most sparrows will hop during the light even if light is very short (e.g., three min of light presented every 10 min; see Chapter 7). However, following a cycle (such as LD1.5:1.5 in Fig. 3.2), a sparrow placed in DD exhibits a circadian rhythm with period near 24h, not a 3h rhythm (Fig. 3.2E). In other words, the sparrow has not learned a 3h rhythm. Direct control by light also occurs in nocturnal animals. For example, Aschoff and Von Goetz [20] describe running in response to dark in hamsters.

When an external cycle, such as a light-dark cycle, synchronizes a biological rhythm, we use the word *entrained* to describe the synchronized biological

Figure 3.1 Sparrows entrain to light-dark cycles. The figure is an event record of a sparrow kept in LD12 : 12 (lights-on 6 a.m. and lights-off 6 p.m.). The bird synchronized with the light-dark cycle, beginning activity at lights-on and ceasing activity at lights-off. The record is 17 days long; each line is 24h of data. A bimodal pattern is not present.

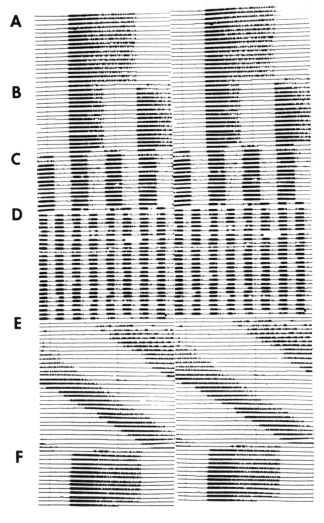

Figure 3.2 Some organisms are directly controlled by light and dark. An example is the event record of a sparrow subjected to a sequence of light-dark cycles including some non-24h regimens. The record has been doubled so that a horizontal line is 48h long.

In (A), the sparrow was in LD12 : 12 (lights-on 6 a.m., lights-off 6 p.m., T = 24).

In (B), the sparrow was in LD6 : 6 (lights-on 6 a.m., 6 p.m.; lights-off, N,M; T = 12) where it was active in the light.

In (C), the sparrow was in LD3 : 3 (T = 6) where it distributed most of its activity to the light.

In (D), the sparrow was in LD1.5 : 1.5 (T = 3) where it was active in the light.

In (E), the sparrow was placed in DD where it freeran. A line drawn through the ends of activity extrapolates to the time of the last lights-off. Sparrows treated with LD1.5 : 1.5 or LL exhibited average phases of onsets of activity in DD (14.9–16.0h after the last lights-off) [53].

rhythm. (Fig. 3.1, 3.2A, 3.2F show *entrainment* to LD12:12.) Clearly (Fig. 3.2), it can be ambiguous for LD12:12 as to whether the cycle is an entrained cycle.

It is possible that both direct control and biological clock control occur simultaneously. It can be frustrating when it is difficult to decide which is occurring. The term *masking* has sometimes been applied to the situation—the biological rhythm is present but is occluded (masked) by the direct effect, or, the direct effect influences the measured parameter without altering the underlying biological rhythm process. To draw conclusions about effects on a rhythm, we usually need to find a way to rule out the possibility of masking.

In considering a Zeitgeber, light in particular, we have to recognize that the signal itself has several parts—dark-to-light transition, lights-on, and light-to-dark transition. Pittendrigh [256] discriminates between two possibilities: (1) a continuous action of light, and (2) light being an effective signal only as it goes on and off. The continuous action idea, also called *parametric entrainment*, is especially compatible with the effects of light intensity on period length (Chapter 2). The transition idea, also called *nonparametric entrainment*, can explain the fact that lights-out determines phase (Figure 5.6). As suggested, arguments can be made for either way for light to act in entrainment, or better yet, a combination of both ideas.

TIME CUES

A German word, *Zeitgeber*, which means *time-giver*, denotes the environmental signals that provide time cues—lighting, temperature, food availability, sound, social factors, etc.

The most well known and most studied of the potential signals for circadian rhythm synchronization is the daily *light-dark cycle* provided by the rotation of the earth. This provides a precise daily signal. Moreover, at most latitudes, the durations of light and dark fluctuate seasonally in a precise way. In the laboratory, the circadian rhythms of most organisms entrain readily to 24h light-dark cycles. The photoreceptor for mammals is the obvious one, the eyes. Preventing light of a light-dark cycle from being perceived by the eyes results in a freerun; for example, Richter's blinded rats freeran in light-dark cycles (Fig. 9.1) [282]. Amazingly, some vertebrate organisms (among the birds, reptiles, amphibians, fish) synchronize to light-dark cycles even when blinded (Fig. 3.3).

The ability to detect light in the absence of the eyes has been called *extra-retinal light perception*. As yet, the site of the extra-retinal light receptor has not been located, but investigators who have used various procedures to prevent light exposure of regions of vertebrate bodies have pointed to the head, especially the brain, as the probable locus [27, 351].

As a result of the daily and seasonal cycles of sunlight, there are less precise daily and seasonal *temperature cycles* that provide potential Zeitgebers. It seems reasonable that some organisms should be able to use the cyclic changes in temperature as time cues. Some organisms can entrain to very low amplitude temperature cycles—less than a degree change per cycle (Fig. 3.4).

Social factors have also been proposed as Zeitgebers. When Menaker and Eskin [231] placed two sparrows in adjacent but separate cages, their circadian rhythms freeran together for a time; moreover, they were able to get sparrows to entrain to cycles of recorded song. Thus, sound can act as a Zeitgeber. We placed pairs of sparrows in common cages and found that even when housed together, many of the pairs exhibited only one freerun, and there

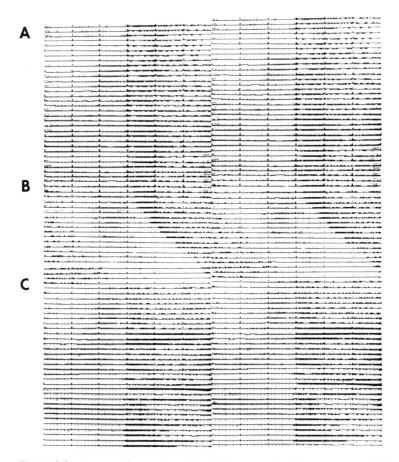

Figure 3.3 Sparrows have extraretinal light perception. The event record of a blind sparrow has been double plotted so that a line is 48h. At (A), the sparrow synchronized with an LD12 : 12 cycle (lights-on noon; lights-off midnight). At (B), the sparrow was placed in DD in which it freeran. At (C), the sparrow was replaced in LD12 : 12 to which it rapidly resynchronized.

were instances of two rhythms in 40–50 percent of the pairs (Figs. 3.5, 3.6) [58]. Male/female pairs of hamsters separated only by a wire barrier "had no effect on the timing of the other's" wheel-running rhythm, according to Davis and co-workers [118]; however, the male of one pair of hamsters exhibited a second activity component matching the circadian rhythm of its female partner. The result, that some paired animals can freerun independently, leads to the astonishing conclusion that it is possible for an organism to run on its own biological clock time in the face of potentially synchronizing time cues from its fellows. Presumably, this is not usually the case. Aschoff and co-workers [18] measured rhythms from four human subjects for four days of constant dark

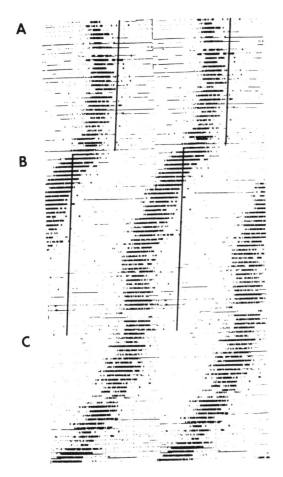

Figure 3.4 Temperature synchronizes the activity of a lizard. The record has been doubled horizontally so that a line is 48h long. In (A), the lizard synchronized with a 0.9°C temperature cycle (18.8–19.7°C). In (B), the lizard followed the cycle when it was advanced 9h. In (C), the lizard freeran in constant temperature (19°C). (After Hoffman [177].)

with rigorous 24h sleep and feeding schedules and learned that their temperature and other rhythms remained synchronized, which they considered evidence for *social* entrainment. Small deviations in freerunning period would probably not have been detected in four days, but since the human average freerun is 25h, it was possible to get a 4h change by the end of four days; therefore, it would appear the individuals were synchronized. Some humans may not be synchronized with their fellows; for example, according to a discussion by Winfree [386], about half of blind folks, and even some people with vision, have 25h freerunning components in their activity/rest cycles.

In the natural world, most of the potential Zeitgebers recur at 24h (light, high temperature, social activity) and should reinforce one another in entraining a circadian rhythm in those organisms that are sensitive to more than one Zeitgeber.

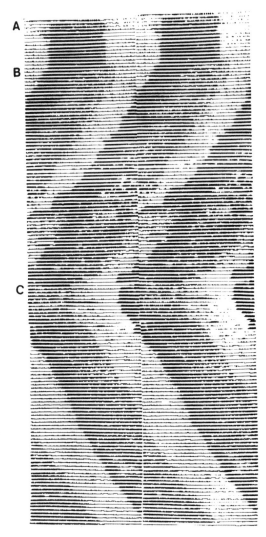

Figure 3.5 Rhythms of pairs of sparrows give clues to social timing interactions. Event record of a pair of sparrows housed together throughout a study. The record has been doubled so that the horizontal axis is 48h.

In (A), the pair was kept in LD12 : 12 to which they entrained, almost as one.

In (B), the pair was placed in LL where they exhibited a freerun with long activity time and with shorter than 24h period length. In the bottom third of the B section of the record, the pair exhibited two rhythms for about two weeks.

At (C), the pair was placed in DD where a single freerun was recorded [58].

LIMITS OF ENTRAINMENT AND FREQUENCY DEMULTIPLICATION

I have said that a circadian rhythm can entrain to a 24h cycle. The length of the cycle providing the time cues is designated by the letter "T". It is also possible for circadian rhythms to entrain to cycles that are not exactly 24h in length. Experiments with exotic cycles of unnatural lengths are sometimes referred to by experimenters as *T experiments*. Sparrows, for example, are active in the light of cycles that are longer or shorter than 24h, though the data are sometimes ambiguous as to whether the sparrows are entrained or whether mask-

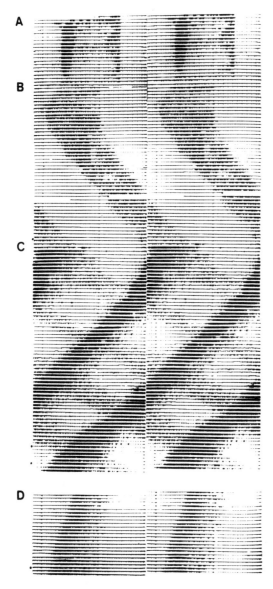

Figure 3.6 Some sparrows may have little response to social time cues. The figure is an event record of a pair of sparrows housed singly or together. The record has been doubled so that the horizontal axis is 48h.

In (A), one of the sparrows was held in LD12 : 12 (lights-on 6 a.m., lights-off, 6 p.m.).

In (B) and (D), the birds were placed in DD (separately) so that their individual rhythms could be recorded.

At (C), the bird from (D) was placed in the cage of the bird whose record is shown in (A) and (B) and recording continued. Two activity components were recorded [58].

ing (direct effect of light) is taking place (Fig. 3.7). However, circadian rhythms generally cannot entrain to all period lengths. Investigators have concluded that there are *limits of entrainment* or a *range of entrainment*. Erwin Bünning described this idea in his book:

> "The free-running periods of about 24 hours are normally *entrained* to exactly 24 hours by 24-hour LD cycles. Plants and animals, however, are limited in their ability to adapt their rhythms to abnormal LD cycles . . . The limits of entrain-

A

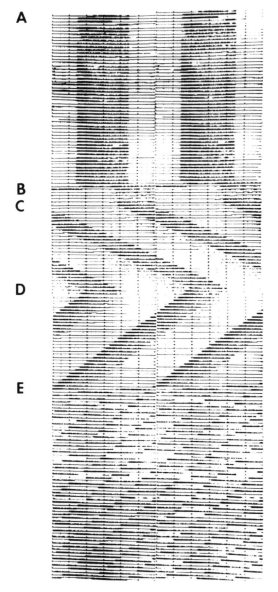

B
C

D

E

Figure 3.7 Organisms can time their activity with non-24h regimens. A record from a sparrow treated with several light-dark cycles of differing length (T). The record is doubled horizontally so that a line of data is 48h long.

In (A), the sparrow was in LD12 : 12 (lights-on 6 a.m.; lights-off 6 p.m.). The length of the cycle was 24h (T = 24).

In (B), there were technical problems.

In (C), the sparrow was in a longer-than-24h cycle (approximately LD8 : 17). The sparrow synchronized with the cycle with a period of 25.6h.

In (D), the sparrow was placed in a shorter-than-24h cycle (approximately LD8 : 15). The sparrow synchronized with the cycle with a period of 23.1h.

In (E), the sparrow was placed in a still shorter cycle (approximately LD8 : 7). The sparrow was active in the light periods of the cycle.

Note that in (C–E) the sparrow was not freerunning, it was in non-24h light-dark cycles in which most of its activity occurred in the light-time. The light-dark cycles are *approximate* because of ±15 min inaccuracy in setting the repeat cycle time (Eagle Signal Co.). (Binkley and Mosher, 1988.)

ment depend upon the special experimental conditions and upon the species. The oscillator can be entrained not only to a 24-hour rhythm, but also in most cases to periods of 22 or 20 hours (for example in an LD11:11 or 10:10). But if the deviations are very large, certainly if one tried to entrain by periods of less than 16 hours, the control proves to be difficult or impossible. The organism then shows its free-running period of about 24 hours. In mice and hamsters this limit may be reached even in cycles of 21 or 26 hours. In other animals, and usually also in plants, the entrainment is still possible down to a period of 18 hours, though great irregularities already occur. Usually the upper limit of plants, animals, and humans is about 28 to 30 hours. In all cases, the limits of entrainment will be reached earlier when working with rather low intensities of light." [90]

For sparrows, Eskin [150] placed the *lower limit* of entrainment between 15.8 and 17.8h because only 25 percent of sparrows entrained to T = 15.8h, whereas 80 percent of sparrows entrained to T = 17.8h. Similarly, the *upper limit* of entrainment was between 28.0h and 28.7h. He used 6h of light per cycle as was done in the experiment shown in Fig. 3.7. Because of masking, it can be difficult to decide whether entrainment is occurring. One way to reduce the impact of potential masking is to use a weaker signal (e.g., dimmer or shorter light exposures). When only 2h of light were used for a cycle in which T = 16h, four kinds of results were obtained: (1) activity in the light, (2) activity in the light with arrhythmia in the dark, (3) activity in the light, synchronization of the activity with some of the D/L to achieve a circadian rhythm, and (4) activity in the light plus a shorter-than-24h freerun (Fig. 3.8). One interpretation of the data is to view the activity in the 2h light as *masking* and the activity in the dark as representing the circadian rhythm (or lack thereof). In this interpretation, only the synchronized bird would be considered entrained (Fig. 3.8C). However, it did not entrain with a period of 16h (the length of the T cycle); rather, it entrained with a period closer to 24h that it was able to extract from the onsets and ends of some of the 2h signals.

 Sometimes, as in the case above or with cycles such as LD6:6, LD3:3, or LD2:2, the rhythm is regulated in the manner in which it performs in LD12:12 [90]. This phenomenon is called *frequency demultiplication* (Chapters 3 and 7). For example, the leaf rhythm of *Canavalia* persisted with a 24h rhythm in LD6:6 (Fig. 3.9). Clearly, frequency demultiplication did not occur in a sparrow exposed to strong LD6:6 signals (Fig. 3.2).

PHOTOPERIOD, SCOTOPERIOD, AND PHASE ANGLES

In a light-dark cycle, the length of the light-time is the *photoperiod* and the length of the dark-time is the *scotoperiod*. Though most authors tend to refer to *photoperiod*, it is worth remembering that in a 24h cycle there is always a scotoperiod that is 24h minus the length of the photoperiod. This becomes especially important in seasonal phenomena where measurement of the night-length may be important (Chapter 6). Entrainment is affected by photoperiod. For example, sparrows entrain to extremes of photoperiod such as 1h or 23h of light (Fig. 3.10) [53]. When the photoperiod was six or more hours, the sparrow was active throughout the photoperiod. The density of activity was not constant; there was often more dense activity just after lights-on and just before lights-off. The bimodal distribution of activity (Chapter 1) and tendency for intense activity just after dawn is common in birds. It is a reason that the morning is the best time for bird watching.

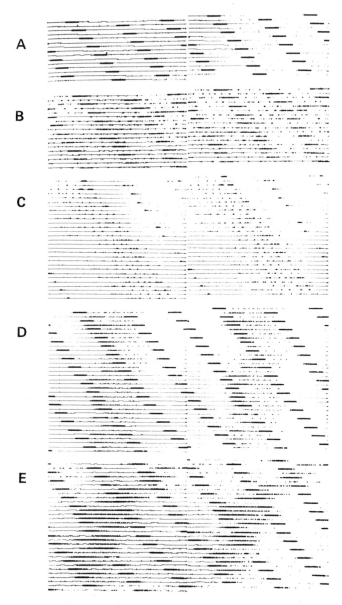

Figure 3.8 Records from five sparrows exposed to LD2 : 14 showing the variety of their responses. (A) was mostly active in the 2h light pulses, (B) exhibited more dark activity, (C) synchronized to alternate 2h pulses as to a skeleton photoperiod, (D) and (E) exhibited shorter-than-24h circadian rhythms in the presence of the 2h pulses in which they exhibited increased activity (an incidence of masking).

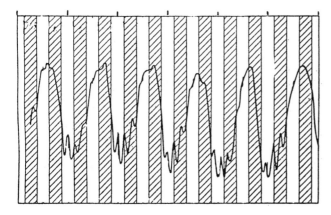

Figure 3.9 Frequency demultiplication in *Canavalia* kept in LD6 : 6. The vertical lines denoted by tic marks above the graph are at 24h intervals; hatching indicates the 6h dark periods. The leaf movements occurred at 24h intervals despite the 12h light cycle. (After Kleinhoonte [201] discussed in Bünning [90].)

But, in less than 6h of light per cycle, the sparrow exhibited activity in the dark which anticipated and trailed lights-on. It is as though there was a lower limit to the number of hours the sparrow could remain inactive. The relationship between the entraining cycle and the sparrow's activity can be quantified by selecting a *phase reference point* in the light-dark cycle (e.g., lights-on), and selecting a *phase reference point* for the sparrow's activity (e.g., onset of activity). The difference between times of the two phase reference points is called the *phase angle*. As can be seen from the sparrow's record in Fig. 3.10, the phase angle is a function of photoperiod—it is greater in LD1:23 (Fig. 3.10C) than in LD4:20 (Fig. 3.10E). We can plot phase angles for the onsets and ends of activity for a number of sparrows exposed to a sequence of light-dark cycles (Fig. 3.11). The figure illustrates another response sparrows have in photoperiods; more sparrows are bimodal when the photoperiod is not extremely long or short.

RE-ENTRAINMENT TO ADVANCED OR DELAYED CYCLES

When we travel east, the sun rises and sets earlier at our destination. In other words, the natural light cycle at the destination is *advanced* with respect to the cycle we were used to at home. Thus, in traveling eastward, we experience a short night followed by a new cycle. When we travel west, the sun rises later than at home, the cycle at the destination is *delayed*, and we experience a long day followed by a new cycle. In other words, we undergo a *phase shift* of the Zeitgeber(s) in our environment. We notice this most when the travel is across time zones and we have to reset our watches. North–south travel does not entail this phase-shifting phenomena, but it may involve a change in photoperiod because of change in latitude. Some journeys would result in changes both in phase and in photoperiod. As a result of travels, the internal circadian

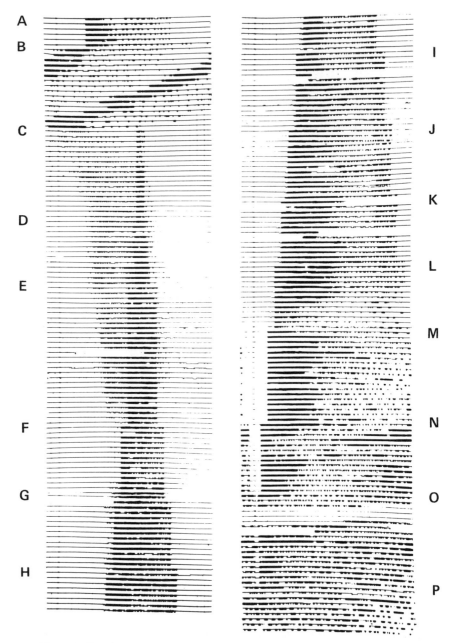

Figure 3.10 The duration of activity in sparrows is a function of the lighting to which they are subjected. The record is from a house sparrow subjected sequentially to a series of different photoperiods. Each line is 24h long; the record continues in the second column at the right. The sequence of treatments was: A = LD12 : 12 (lights-on 6 a.m. and lights-off 6 p.m.); B = dim LL (240 lux);

rhythm must reset to regain synchronization with the Zeitgebers in the environment.

We can easily mimic this in the laboratory by subjecting experimental organisms to phase shifts of their entraining cycles. When we do this, three observations can be made. First, the organism adapts to the new cycle. Second, often there are intermediate cycles as the adjustment is made. Third, the rate of resetting is dependent upon the strength of the Zeitgeber. For example, when a sparrow kept in dim LD12:12 (240 lux) was phase shifted by delaying the times of lights-off and lights-on for 6h, it reset; intermediate cycles were seen for three days (Fig. 3.12).

The word *transients* has been defined as "temporary oscillatory states between two steady states" [9] and it applies here; the two steady states are the entrainment in the pre- and post-shift cycles, the transients are the intermediate cycles. When transients are seen, they are evidence of the circadian oscillator. In that sense, transients are proof of entrainment and evidence against masking.

Circadian biologists use conventions for denoting phase shifts. They set 24h = 360 degrees and refer to phase shifts in degrees; further, they denote delaying phase shifts by minus, and advancing phase shifts as plus (e.g., the 6h delay in Fig. 3.12 would be a -90 degree phase shift). The phrase *phase reversal* is sometimes used to denote a 180 degree phase shift, reversing the times of light and dark (Table 3.1).

The rate at which readaptation occurs is partly a function of the conditions used during the phase shift. For example, sparrows reentrained to a 180 degree phase shift of LD12:12 (240 lux) in 1.4 +/− 0.4 days if the light was extended to make the phase shift, but took 5.0 +/− 0.8 days if the dark was extended for 24h to make the phase shift [36].

The interval of readaptation during travel is generally referred to as *jet lag*, defined by Ehret and Scanlon as the "transient state of dischronism" [141]. Travel is not the only area of human endeavor where the rate of phase

---

C = LD1 : 23;   D = LD2 : 22;   E = LD4 : 20;   F = LD6 : 18;   G = LD8 : 16; H = LD10 : 14;   I = LD12 : 12;   J = LD14 : 10;   K = LD16 : 8;   L = LD18 : 6; M = LD20 : 4; N = LD22 : 2; O = LD23 : 1; P = bright LL (870 lux). Comparing (B) and (P) illustrates the effect of constant light intensity on sparrow rhythms—the sparrows exhibit short period rhythms in dim LL but arrhythmic behavior in bright LL.

Heavy activity occurred during the light-time on all the cycles (A, C–O) and, in fact, it is possible to determine when the lights were on merely by looking at the bird's record. However, activity in the dark-time also occurred, particularly when the duration of light was shorter than 10h (C–G). That is, the bird's activity anticipated lights-on and the amount of the anticipation was dependent upon the photoperiod. This bird appeared to be synchronizing the end of its activity with lights-off, but some birds exhibited activity both before and after the short light period [36, 201].

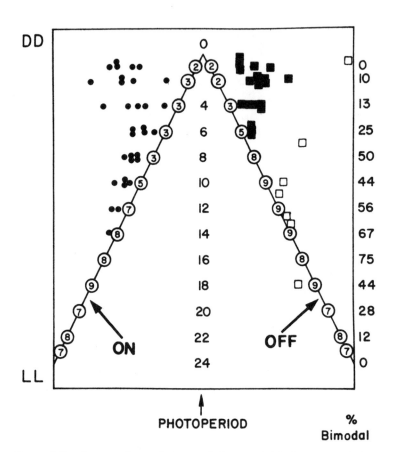

Figure 3.11 Phase angle is a function of photoperiod [49]. The graph shows activity onsets and ends in animals plotted with respect to lights-on (ON) and lights-off (OFF) times in different photoperiods. The horizontal axis is 24h and represents the photoperiod; the vertical axis represents different treatments ranging from constant dark (top) through short to long photoperiods and constant light (bottom). % *bimodal* is the percentage of sparrows that displayed bimodal activity patterns in the corresponding photoperiod.

Sparrow activity onsets are represented by the filled circles and the numbers in the open circles superimposed on the ON line (these numbers are the numbers of sparrows that began activity at the time of lights-on).

Sparrow activity ends are represented by the filled squares and the numbers in the open circles on the OFF line (these numbers are the numbers of sparrows that ceased activity at the time of lights-off).

Open squares represent activity onsets of golden hamsters in various photoperiods. The hamsters, of course, are nocturnal and begin activity in the dark-time. The hamster data were in Elliott [142].

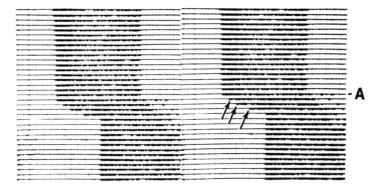

Figure 3.12 A record of a house sparrow's perching activity. The record is double plotted so that the horizontal axis is 48h and the vertical axis is time in days. The sparrow was active in the light of the LD12 : 12 to which it was exposed. At (A), the phase of the light-dark cycle was delayed by 6h and the sparrow exhibited three days of transients as it delayed to regain synchrony with the cycle.

TABLE 3.1 Rates of Phase Shifting[1]

Organism	Days
Human	11
Rat	5–8
Mouse	9
Crab	6
Chicken	3–7
Sparrow	4
Roach	4

[1]Number of days required for resynchronization after a 12h phase shift (= phase reversal, = 180 degree phase shift) after data of Aschoff [19]. Rates of phase shifting vary depending on the rhythm measured and the nature of the Zeitgeber used.

shifting is important. In order to provide 24h services or make 24h use of expensive equipment, many industries have made a practice of using various shift work schedules to obtain a continuous labor force. Often the schedules require frequent resynchronization of individual workers. Obviously, it is desirable to minimize the consequences of time shifting in travel or shift work (Chapter 10).

SKELETONS

It is possible to mimic some effects of light-dark cycles with pulses of light. The term *skeleton photoperiod* is defined by Saunders as a light regimen using two shorter pulses of light to simulate dawn and dusk effects of a longer complete photoperiod [303]. When the duration of the two pulses is equal, the skeleton photoperiod is said to be *symmetrical;* when the duration of the two pulses is different, the skeleton photoperiod is said to be *asymmetrical.* Skeleton photoperiods have been especially useful for investigating photoperiodic phenomena (Chapter 6). Pittendrigh considers skeleton photoperiods to be a "special case of frequency demultiplication" [266].

The effects of one symmetrical skeleton photoperiod on a circadian rhythm can be seen in a locomotor record of a sparrow (Fig. 3.13). The skeleton consisted of 1h of light alternating with 11h of dark, LD1:11. This is the skeleton of LD13:11. How the sparrow entrains to the LD1:11 seems at first ambiguous since the sparrow can choose to be active in either dark period and still be in the skeleton of LD13:11. In experiments, the choice is not random but depended on prior history. When the time of first presentation of LD1:11 was varied following the last L/D (lights-out) of the prior LD12:12, the birds synchronized the onset of their activity with the pulse that occurred 5–18h after the time of the last L/D. Eighty-seven to 94 percent of the sparrows delayed when the first light pulse fell in the late-day and early subjective night and there were no transients in 35 percent of the sparrows that advanced if the light pulse fell in the late-night and early-day [53, 329]. The results correspond to predictions based on the sparrows' phase response curve (Chapter 4). Note that up to eight transients are visible (Fig. 3.13D) when the sparrow was shifted from LD12:12 to LD1:11, whereas at most three transients were present (Fig. 3.13G) when the sparrows were shifted from the skeleton back to LD12:12; presumably this is partly because the skeleton is a weaker Zeitgeber than the full photoperiod.

It follows that there are skeleton photoperiods that can be interpreted two ways by the organism. The *bistability* phenomenon is the fact that there is a

"narrow range of skeleton photoperiods (10.3 < 13.7) in *D. pseudoobscura*, where the two dark intervals are close to tau/2, and within which two different steady states are possible for any zeitgeber." [260]

Presumably this is the case for LD1:11.

The effects of another skeleton photoperiod upon a circadian rhythm can also be illustrated with sparrows. We subjected sparrows to LD1:6:1:16. From the investigators' point of view, this cycle could be the skeleton of LD8:16 (a short photoperiod) or of LD18:6 (a long photoperiod). However, the sparrows were unanimous in entraining to the cycle as though it were the skeleton of LD8:16 (Fig. 3.14). It is possible to interpret this result with the sparrows'

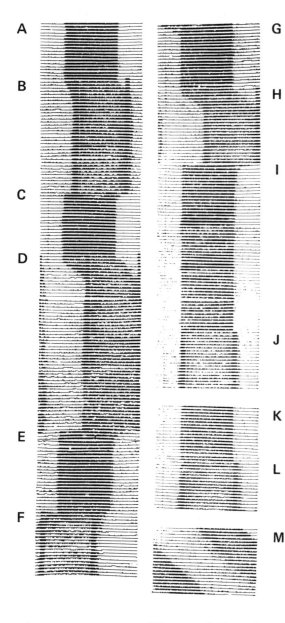

A

B

C

D

E

F

G

H

I

J

K

L

M

Figure 3.13 Sparrow synchronization to skeleton cycles depends on
the timing of the pulses with respect
to the sparrow's subjective time. The
figure is an event record of a sparrow
treated with LD12 : 12 (lights-on 6
a.m., lights-off 6 p.m.) and a skeleton
cycle, LD1 : 11. The lines are 24h
long and the record continues in the
second column. In this case, the skeleton was unambiguous, the cycle
was the skeleton of LD13 : 11. The
sparrow synchronized with the skeleton in about a week but took less
time to reentrain to LD12 : 12.
       The sparrows in the study syn-
chronized the onset of their activity
in LD1 : 11 to the pulse that occurred 5–18h after the time of the
last lights-off of the LD12 : 12 pretreatments. The sparrows synchronized with the pulse that
required the smaller of the alternative phase shifts by which they could
resynchronize; we called this conser-
vation of phase [53].

phase response curve (Chapter 4). In such an interpretation, a second pulse 18h
after D/L (lights-on) would always fall in the sensitive middle of the sparrows'
phase response curve, causing them to phase shift; whereas a pulse 8h after
D/L would fall in a *dead zone* where less phase shift would result. Thus, the
reasoning is that the LD8:16 interpretation is favored because it is more stable
[59]. To paraphrase Pittendrigh [260], there is a *minimum tolerable night length*

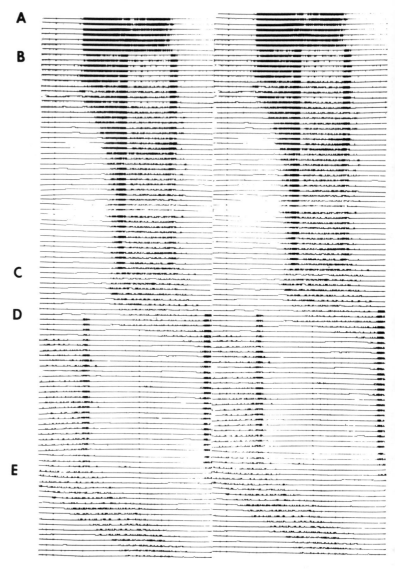

Figure 3.14 Sparrows entrain to *skeleton* light-dark cycles in which dawn and dusk are represented by light pulses. The figure is an event record of sparrow activity (double plotted, horizontal axis = 48h) kept in the skeleton photoperiod LDLD1 : 6 : 1 : 16. The cycle could be interpreted as the skeleton of LD8 : 16 or as the skeleton of LD18 : 6. The sparrows invariably selected the LD8 : 16 interpretation even when the cycle was phase shifted.

In (A), the sparrow was in LD12 : 12 (lights-on 6 a.m., lights-off 6 p.m.). (B) was the first presentation of the skeleton—the bird took almost three weeks to entrain to the cycle during which transient activity anticipating lights-on of one pulse is present. At (C), the sparrow was placed in DD in which it freeran from a phase setting clearly derived from the prior entrainment to the skeleton. At (D), the skeleton was reintroduced but with the pulses occurring at different times than in (B); the sparrow took about a week to synchronize with the skeleton. In (E), the sparrow was again placed in DD, where it exhibited a freerun [59].

(about 10.3h in *Drosophila*); when the interval is less, then the pacemaker makes a *phase jump* so that its subjective night falls in the longer dark period.

INCOMPLETE ENTRAINMENT

Sometimes a signal is right at the threshold for entrainment—strong enough to *catch* a rhythm but too weak to *hold it*. There are a number of patterns that have been observed and they have been given various names: *relative coordination* (*failing* or *incomplete* entrainment), *bounce*, and *breakaways*.

In relative coordination, the organisms' rhythm remains roughly synchronized by the cycle, but the rhythm drifts away from the signal and then periodically regains it. Relative coordination has also been described as freerun in which there is "a recurring impact of the zeitgeber on the rhythm . . . visible as a regular beat phenomenon"—an "oscillatory freerun" [266]. Relative coordination was observed, for example, in golden hamsters using 15 min light pulses as Zeitgebers in T cycle experiments [266]. In relative coordination, the rhythm appears to freerun across the signal, then catch it for awhile, then freerun for awhile, and so on. The freerun portion is in the same direction.

Sometimes, however, the organism freeruns past a Zeitgeber with a long period, then reverses direction to freerun with a short period to catch the signal. This behavior has been called bounce and has been reported for mice exposed to 24h cycles with only 15 min of light per cycle and in a mouse exposed to a 3h skeleton photoperiod [266].

A Siberian hamster in our laboratory gave us an example of incomplete entrainment in LD12:12 (Fig. 3.15). There was an oscillation in the onset of the hamsters' activity in the dark-time.

The word "breakaway" has been used to denote the times when a rhythm freeruns in the presence of a signal to which it sometimes is synchronized. A possible example of a breakaway occurs in Fig. 3.5B.

MULTIPLE ZEITGEBERS

There are several cases of multiple Zeitgebers to consider. First, in the natural environment there are cycles with a common period that provide reinforcing signals to an organism. For example, for the daily cycle, light and high temperature usually coincide. Second, an organism in the natural environment is subject to potential external Zeitgebers for more than one rhythm at a time; for example, a 24h cycle from the sun, a 12.4h cycle from the moon, an annual cycle of photoperiod, etc. Third, there is more than one internal rhythm; for example, a woman experiences menstrual as well as daily cycles. Thus, it is of interest to examine the *interaction* of rhythms and the effects of two competing signals (*conflicting Zeitgebers*) upon a rhythm.

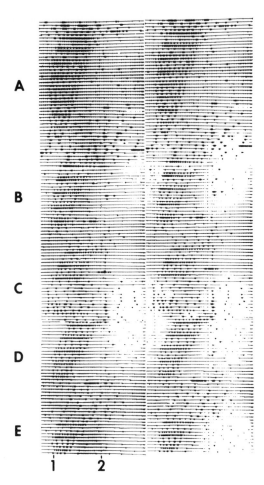

Figure 3.15 Not all animals entrain clearly in all light-dark cycles. Record of a Siberian hamster kept in LD12 : 12 (lights-on at 1, lights-off at 2); the record has been doubled horizontally so that a line represents 48h of data. The hamster appeared to make five attempts to entrain to the cycle (marked by the letters A–E). (Binkley, Spangler, and Mosher, unpublished data.)

Superposition can be obtained when an exogenous cycle forces its period in a physiological variable that also exhibits a circadian rhythm (Figure 3.16). Superposition occurred, for example, in the circadian rhythm of mobility in *Euglena* in LD2:2 [90] and in a sparrow kept in a 3h cycle (Fig. 3.17D).

Classic studies of conflicting Zeitgebers have involved subjecting poikilotherms (so called cold-blooded animals) to simultaneous temperature and light-dark cycles [256]. In studies of *Drosophila* and cockroaches, it was found that the *Drosophila* eclosion rhythm would follow the low point of a temperature cycle throughout the light-time. Similarly, the phase of a cockroach followed the high point of a temperature cycle during the dark-time [288]. However, the *Drosophila* rhythm could not follow the temperature cycle in the dark and the roach rhythm could not follow it in the light. There was a 180 degree *phase jump* that occurred at a *critical phase angle*. There was a *zone of forbidden phase* in which the temperature cycle could not drive the rhythms.

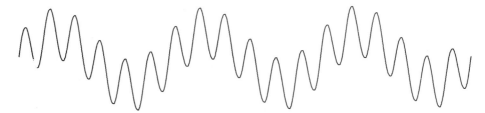

Figure 3.16 The figure illustrates *superposition* in which one sine wave with a period of 24h was added to a sine wave with a period of 4h.

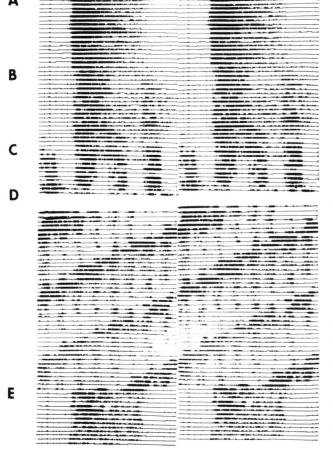

Figure 3.17 Short periodicities can be forced with light-dark schedules in some species even if they do not normally organize their activity into short bouts. The record shows the response of a sparrow to a sequence of light-dark cycles. In (A), the sparrow was in LD12 : 12 (lights-on 6 a.m.). In (B), the sparrow was exposed to LD6 : 6 in which it was most active in the 6 a.m. to noon light; this is the only example we were able to obtain of possible frequency demultiplication in a house sparrow. In (C), the bird was active in the light of LD3 : 3. In (D), the sparrow was placed in LD1.5 : 1.5; there was activity in many of the 90 min light periods, but the bulk of the activity freeran with a period shorter than 24h. The bird re-entrained to LD12 : 12 in (E). The amazing thing about this record is that the sparrow from which it was obtained was blind.

A number of investigators have exposed rodents to simultaneous light-dark and restricted feeding schedules. Edmonds and Adler [130, 131] used rats as subjects. When rats, which are normally nocturnal and feed in the dark, kept in LD12:12 or LL were fed only for 2h in the light, the rats showed a burst of wheel running in anticipation of the food presentation. They concluded that timed meals were as effective a Zeitgeber for rats as lighting. When, in another experiment, they presented rats with two simultaneous food cycles—a 24h and 25h cycle—they obtained multiple results; the rats synchronized to one cycle, both cycles, or had activity freerunning between the two cycles. They concluded that their data supported the hypothesis that rats were in possession of multiple, and separable, circadian oscillators.

4

RESETTING
THE
BIOLOGICAL CLOCK

"There are several possible ways of obtaining data for response curves. Type I . . . (uses) free running rhythms . . . in a constant dark environment . . . interrupted by light signals . . . in type II . . . the light pulse is applied in DD, but is preceded by entrainment to a light-dark cycle . . . Type III utilizes a light step . . . in type IV, a free running rhythm crosses a daily repeated light-pulse which is not strong enough to be a Zeitgeber . . . . Type V consists of changes from LD to DD with the last transition from L to D being shifted relative to its normal position . . . in type VI, a light pulse is applied in the dark-time of LD."

Aschoff [10]

RESETTING

I have already introduced some information concerning the resetting of the biological clock (see Chapter 3). In that section I considered how the biological clock resets when the phase of the entraining cycle is shifted (as occurs during east–west travel or shift work). But the resetting mechanism is even more basic to the functioning of the biological clock. At most latitudes, organisms face a change in Zeitgeber timing with each new day's dawning. Because an organism's freerunning period is not 24h, the organism must reset every day or at frequent intervals if it is to stay in synchrony with the natural day-night cycle.

A tool that has been used extensively to study resetting has been the measurement of phase shifts following single short pulses (*perturbations*) of a Zeitgeber. The bulk of experiments have used short light pulses (e.g., 10 min to 6h), but some investigations have also used dark pulses, temperature pulses, steps from light-to-dark, etc. Just as when the phase of an entraining cycle is shifted, following a perturbation of the Zeitgeber, we may see *transient* cycles before a new *steady state* is attained. In early experiments, short light signals delayed or advanced the circadian rhythms of leaf movements in *Canavalia* [201], eclosion of *Drosophila*, luminescence of *Gonyaulax*, discharge of sporangia by a *Pilobolus* fungus, and the rhythm of color change of the fiddler crab *Uca* [90]. An example of a sparrow's response to a 4h light pulse appears in Fig. 4.1.

USING PHASE RESPONSE CURVES TO STUDY RESETTING

When scientists methodically exposed organisms to single perturbing signals (e.g., light pulses) at different times over 24h, they found that the phase shifts they obtained were not all the same. It was possible to graph the direction and amount of shift obtained versus the time (usually the midpoint) that the pulse was given. Such a graph is called a *phase response curve* or PRC. The graphs reveal the time course of sensitivity of the circadian rhythm to the signal. Usually the sensitive period occurs in the *subjective* night (*projected dark*, the time the organism could normally expect to be night, e.g., the rest period for a sparrow). We obtained sparrow phase response curves using 4h light pulses on sparrows kept in constant dark following LD12:12 (Figs. 4.2, 4.3). The sparrow PRC has several features that are common in general. Delays occur in the early half of the projected night encompassing the time of expected dusk, advances occur in the late projected night and include the time of expected dawn, and there are times of apparent relative insensitivity around midnight (*singularity*) and midday (*dead zone*).

It is also possible to obtain a phase response curve in sparrows and in other species by imposing dark pulses upon animals kept in constant light following LD12:12. The protocol and phase response curve for sparrows ex-

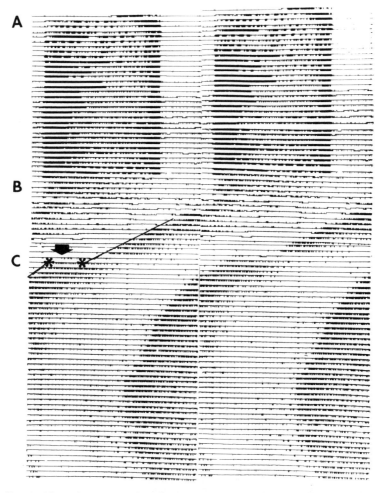

Figure 4.1 Pulses imposed on freerunning organisms phase shift the circadian rhythm. Here, in (A) a sparrow was entrained to LD16 : 8 (lights-on 2 a.m., lights-off 6 p.m.). At (B), the sparrow was placed in constant dark in which it freeran. At (C), a single 4h light pulse was imposed (beginning at the time of the arrow). The diagonal lines trace activity onsets of the sparrow before and after the pulse. The phase shift (about 3.5h advance, Δ phi = +3.5h) was measured on the first day after the pulse as the horizontal difference between the asterisks. The sparrow's record has been doubled; each line is 48h long. The sparrow in this record exhibits a spontaneous increase in period length (a *knee*) in the bottom fourth of the record; such knees are characteristic of the freeruns of about half of sparrows. Another example of a phase shift measured similarly appears in Fig. 1.11.

LIGHT   DARK

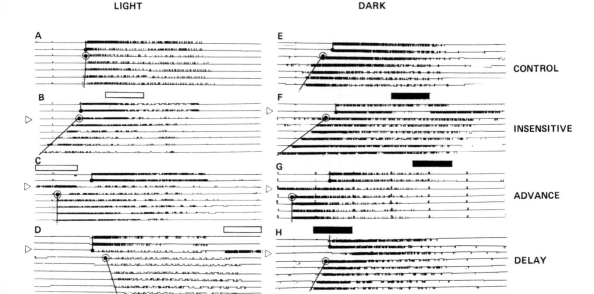

Figure 4.2   Light and dark pulses presented in the first 24h after a light-dark cycle shift the phase of the ensuing circadian rhythm. In the eight examples shown here, sparrows were first entrained to LD12 : 12; the last two days of LD12 : 12 are shown at the top of each record. Controls were placed in constant dark (A) or constant light (E). Single 4h light pulses were imposed on sparrows kept in the dark (left, B–D) and single 4h dark pulses were imposed on sparrows kept in constant light (F–H). The vertical lines and points indicate the onsets of activity before and after the pulses. The *phase shift* was measured as the horizontal difference (in h) between the filled circle (pretreatment phase in LD12 : 12) and the circled filled point (posttreatment phase). Control phase shifts were subtracted from the phase shifts following pulses to obtain the phase shift due to the pulse. For example, (C) and (G) show large advance phase shifts due to a light pulse in the late dark-time (open bar) or a dark pulse at the end of the light-time (filled bar). Usually the birds were active during a light pulse (e.g., in C and D) and inactive during a dark pulse (e.g., in F), a phenomenon that has been called masking. Measurements from experiments such as these can be used to plot a 24h curve of phase shifting, a phase response curve (PRC) [200] as in Fig. 4.3.

posed to 4h dark pulses is shown in Fig. 4.4. Again, the curve has generalizable features. Advances were seen at the time of expected dusk and on into the early subjective night, delays occur around the time of expected dawn, and there are relatively insensitive times. If you compare the PRCs obtained by light and dark pulses, you can see that they are roughly opposite or reversed; some authors describe this by saying that dark and light pulse PRCs are *mirror images* (Fig. 4.5).

It is widely held that phase response curves represent the sensitivity of the underlying circadian pacemaking process. For example, DeCoursey, referring to mammals, says:

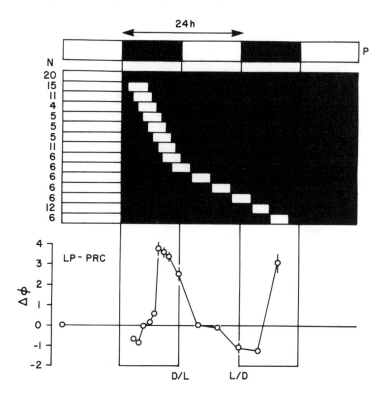

Figure 4.3 Phase shifts (vertical axis) in response to single pulses can be graphed versus time of day of the pulse (horizontal axis); a 24h plot of such shifts constitutes a phase response curve. Here, the graph shows sparrow phase shifts measured by the method shown in Fig. 4.2. D/L = time of projected lights-on and L/D = time of projected lights-off in the pretreatment LD12 : 12 cycle represented by the bar at the top marked P. The column of numbers marked N is the number of phase shifts measured using the 4h light-pulse represented by the bar to the right of the number. The point on the phase shift curve below is plotted at the time of the midpoint of the pulse. Thus the curve represents the sensitivity to 4h light pulses at various times of subjective day and night, and moreover, the direction (advance, positive values; delay, negative values). The sparrows' phase response curve is typical—pulses in the early projected dark-time cause delays, pulses in the late projected light-time cause advances, and there are *dead zones* (where pulses have little effect) in the middle of the projected light and dark times [200].

"Plots of responsiveness to light resetting (phase response curves, or PRCs) are a manifestation of the cyclic function of the neural pacemaker." [122]

However, it should be pointed out that there is another explanation for phase response curves, to wit, that the variation in sensitivity revealed by the PRC represents the rhythm of sensitivity in detection of the Zeitgeber (e.g., the photoreceptor in the case of light pulses) [90].

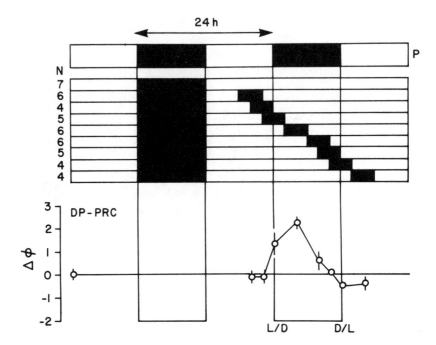

Figure 4.4 Phase shifts (vertical axis) in response to single pulses can be graphed versus time of day of the pulse (horizontal axis); a 24h plot of such shifts constitutes a phase response curve. Here, the graph shows sparrow phase shifts measured by the method shown in Fig. 4.1. D/L = time of projected lights-on and L/D = time of projected lights-off in the pretreatment LD12 : 12 cycle represented by the bar at the top marked P. The column of numbers marked N is the number of phase shifts measured using the 4h dark-pulse represented by the bar to the right of the number. The point on the phase shift curve below is plotted at the time of the midpoint of the pulse. Thus the curve represents the sensitivity to 4h light pulses at various times of subjective day and night, and moreover, the direction (advance, positive values; delay, negative values). The sparrows' phase response curve is typical—pulses in the early projected dark-time cause advances, pulses in the late projected light-time cause delays, and there are *dead zones* (where pulses have little effect) in the projected light and dark times [200].

WAYS OF OBTAINING AND REPRESENTING PHASE RESPONSE CURVES

Scientists have obtained phase response curves for circadian rhythms using a variety of techniques. Aschoff [10] identified six ways to obtain data for phase response curves: (1) Application of a light pulse on an organism freerunning in DD (as for the sparrow in Fig. 4.1). (2) Application of a light pulse in DD preceded by entrainment to a light-dark cycle (the method used for the sparrows in Fig. 4.2). (3) Application of a light-step. (4) Calculation from a freerunning rhythm as it is perturbed by a daily weak signal. (5) Calculation from the shift resulting from L/D transitions (lights-offs). (6) Application of a light-pulse

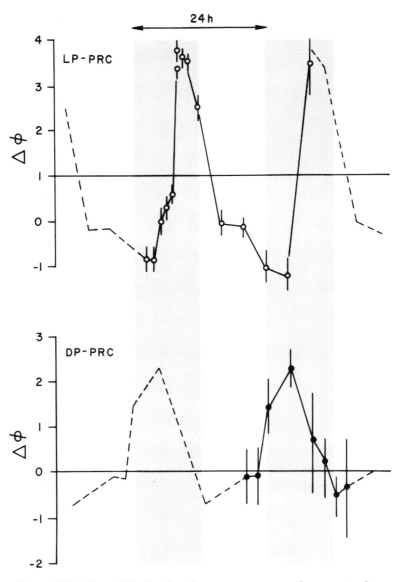

Figure 4.5 Light and dark pulse phase response curves for sparrows from Fig. 4.3 and 4.4 are shown plotted here together to illustrate their phase relationships. Notice that the curves are phase shifted with respect to one another with advances occurring at subjective dusk for dark pulses but before subjective dawn for light pulses [49, 200].

in the dark-time of LD. To this I add (7) Exposure of the organisms to dark pulses in constant light. In all of these methods except (6) and (7), "either the first onset in DD or the extrapolated line through onsets can be used" [10] to measure the phase. "The phase shift is measured as the time difference between the last onset, or average of onsets, before the signal and the onset after the signal" [10] (see Fig. 1.3). When transients occur after a pulse, it is appropriate to measure the phase after the transients have subsided; in any case, the method of determination should be, and usually is, specified. For rhythms (such as the temperature rhythm) for which there is no onset (as there is for locomotor activity), a criterion must be set up to measure phase. The phrase *phase reference point* is used to denote the characteristic of the rhythm that is used to measure phase (e.g., onset of activity in hamsters or sparrows, median of the eclosion peak in *Drosophila*).

Various authors have used different techniques for plotting the phase responses. Usually the phase responses are plotted on the ordinate (vertical axis), and the time of the perturbation (e.g., midpoint of a light pulse) is plotted on the abscissa (horizontal axis). Time units on the abscissa are usually circadian time (Chapter 2). By the convention used in mathematical and physical notations, delaying phase shifts are assigned a minus value; advancing phase shifts are assigned a plus value (sometimes the reverse appears in biology). Sometimes some work is required to compare phase response curves where investigators have used different plotting methods. The problem may be exacerbated if it is not readily apparent whether the organism is nocturnal or diurnal (as with a plant) so that the relationship of hour zero to natural lighting can be determined.

The apparent midnight discontinuity in the phase response curve can be eliminated by plotting the data in another manner (Fig. 4.6). This appears to eliminate the concept of advancing or delaying in a phase shift—after all, a 6h advance can be viewed as an 18h delay. In practice, however, transients or the phase of the first shifted cycle reveal whether the overt rhythm that was measured advanced or delayed. Plotting the PRC as a continuous curve (on the diagonal) does suggest the view that in phase shifting, the organism has simply slipped to a new position on an underlying temporal process.

Winfree and others [100, 386] have measured phase response curves varying the nature of the signal (e.g., duration of light). In doing this they obtain three parameters—time of the pulse, duration of the pulse, phase shift—and so can plot phase response curves in three dimensions as a *circadian surface*. The resulting graph gives the impression of a clock spiraling upward through time.

FACTORS THAT MODIFY PHASE RESPONSE CURVES

As soon as investigators began to measure PRCs, they began to compare the results they obtained using the different methods of obtaining PRCs and the PRCs obtained for different species.

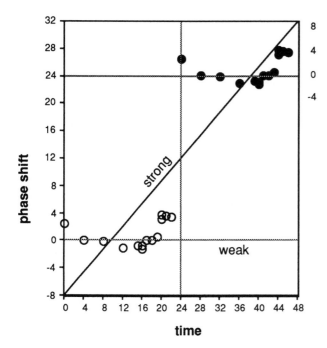

Figure 4.6 It is possible to view phase shifts so that the sign disappears. We can view advances as long delays or vice versa and plot phase shifts to remove the discontinuity that otherwise appears in the midnight [383]. Here, sparrow phase shifts have been double plotted in both the horizontal and vertical directions. The data are close to the horizontal lines through zero because the small signals produced weak resetting. If larger signals were used, the phase shifts would be larger and would fall closer to the diagonal line.

First, as might be expected, PRCs have individual species differences in the magnitude of phase shifts obtained in response to similar signals.

Second, nocturnal or diurnal physiology seems important. It appears that larger phase shifts can be gained with smaller signals in nocturnal animals (e.g., hamsters and flying squirrels respond to 15 min light pulses), whereas larger signals may be required for diurnal species (e.g., sparrows respond to 4h pulses). Another generalization that emerged [256] was that PRCs for nocturnal species should have a larger delaying portion in their PRCs and diurnal species should have a larger advancing portion in their PRCs in order to permit correction of their respectively short and long taus to 24h in entrainment. In the initial view of this, the advance section of a diurnal animal included the time of expected dawn and the delay section of a nocturnal animal (e.g., *Glaucomys*, the flying squirrel) included the time of projected dusk [256]. In practice, however, as for *Drosophila* and *Passer* (sparrows), peaks of PRCs often encompass both dawn and dusk.

Third, a factor that was noted to affect the characteristics of a PRC within a species was the quality of the signal. For example, *Drosophila* PRCs were measured using pulses with 12h, 4h, and $1/2000$ sec (!) duration (Fig. 4.7). The PRCs for the different signals varied in amplitude, and in the phase at which the shifts changed from delays to advances. The effects of signal duration (e.g., of dim blue light from 15–120 seconds) upon the PRC were studied methodi-

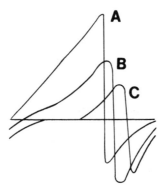

Figure 4.7 Phase response curve shape depends upon the nature of the signal. The graphs are phase response curves for fruit flies in which (A) was obtained with 12h pulses, (B) with 4h pulses, and (C) with $1/2000$ sec pulses. (After Pittendrigh [256].)

cally in *Drosophila* [383]. When the average slope of the resetting curves was calculated for the form of plotting used by Winfree, the resetting was either weak (type 1, odd resetting, using a sufficiently short pulse, weak stimulus) or strong (type 0, even resetting, using a long pulse, strong stimulus) (Fig. 4.6). In other words, as stimulus size increases, the response curve changes from type 1 to type 0.

Fourth, pretreatment photoperiod determines the exact shape of the PRC [55, 264]. In sparrows, for example, the PRC obtained after LD16:8 has a large advance portion, while the PRC obtained after LD8:16 has a substantial delay portion in the early projected night and the advancing portion is reduced (Fig. 4.8). Similarly, hamsters pretreated with LD18:6 had a PRC with only 31 percent of the amplitude of the PRC attained by hamsters pretreated with LD10:14.

Fifth, pretreatment T cycle may affect the PRC. After T = 23h (LD1:22.3), the advancing portion of the hamster PRC is reduced compared with the PRC for hamsters after T = 24h (LD14:10) [260].

In sum, it is clear that the PRC is malleable and is a function of the species, the pretreatment conditions, and the nature of the signal used to obtain the PRC.

INSTANTANEOUS RESETTING

The basic view of *transient* cycles seen during a phase shift is that they represent the observed, measurable oscillation (slave or driven oscillation) as it gradually regains phase with an underlying internal pacemaker oscillation (the biological clock). However, early in the field, it was recognized that models for entrainment required the validity of two assumptions. As Pittendrigh described them:

> "First is the assumption that the response curve based on the steady-state phase-shift is a proper measure of the phase shift executed by a distinct driving oscilla-

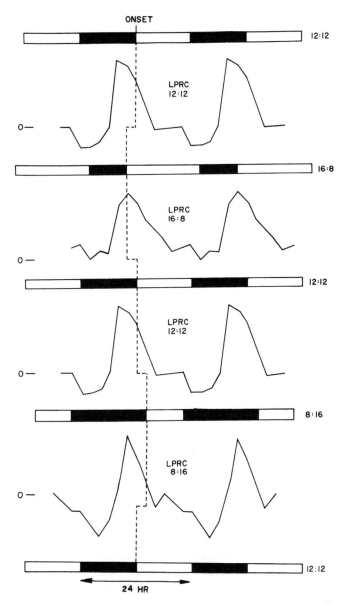

Figure 4.8 Phase response curves depend upon the signal given (e.g., duration of pulse, intensity of light, photoperiod). Here, phase response curves are shown for sparrows. The curves were obtained in the manner of Fig. 4.2 and plotted in the manner of Fig. 4.3. However, the pretreatment cycle shown above each graph was different. The dotted line shows the change in time of onset that the sparrows had to make to adjust to a new photoperiod. The vertical axis is *phase shift* with advances plotted up and delays plotted down. The horizontal axis is 48h; the PRCs have been plotted twice horizontally to aid in their comparison. The figure is intended to show the adjustments that would be required in onset of activity (dotted line) as the seasons progressed from the spring equinox (top) through the summer solstice (LD16 : 8) through the fall equinox (LD12 : 12) to the winter solstice (LD8 : 16) and so forth [55].

79

tion on the day (day 1) the signal is seen. Second is the assumption—essential to computation of the interaction of two signals separated by a short interval within one cycle—that the phase-shift effected by a light signal is completed nearly instantaneously." [258]

In order to study the properties of the pacemaker itself, Pittendrigh devised a protocol involving exposure of organisms to two pulses.

"A first pulse is given at ct 16; second *tester* pulses are used to track the time course of the pacemaker after the first pulse. It is well predicted by the assumption of an instantaneous 6-hr delay of the PRC . . . The same protocol shows that a pulse at ct 20.5 causes an instantaneous 6-h advance." [260]

In other words, the results support the idea that the second pulse falls upon a phase shifting sensitivity curve that was already phase shifted by the first pulse. We tried the same protocol on house sparrows with similar results (Figs. 4.9, 4.10, 4.11). When the sparrows were first given an advancing or delaying pulse, the PRCs due to series of second *tester* pulses were phase shifted. But when the PRCs were graphically compensated for the predicted phase shifting effects of the first pulses (based on the single pulse PRC), most of the points fell on the single pulse PRC, supporting the idea that the first pulse shifted the resetting sensitivity at once.

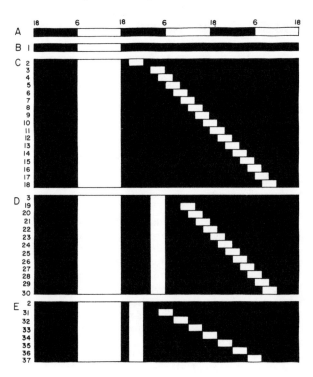

Figure 4.9 Two pulses produce phase shifts that can be predicted based on single pulse response curves assuming the first pulse shifts the response curve so that the second pulse falls on a phase-shifted response curve. The diagram shows the protocol for double pulse experiments in sparrows if an (advancing) first pulse was imposed in the late projected dark-time (D) and a (delaying) first pulse was imposed in the early projected dark time (E). (A) shows the projected pretreatment LD12 : 12 cycle (lights-on at 6h, lights-off at 18h), (B) shows the DD control regimen, and (C) shows the control single pulse protocol [56].

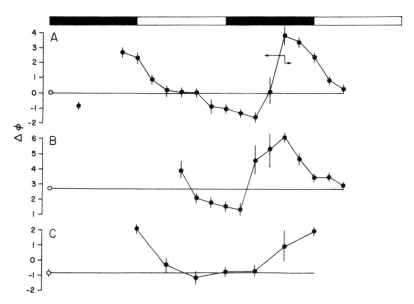

Figure 4.10 Responses obtained with the protocol shown in Fig. 4.9 for studying the effect of two pulses. The bar over the graphs shows the projected LD12 : 12 cycle; (A) shows the results with single pulses (Fig. 4.9, protocol C); (B) shows the two-pulse response curve in which the whole response curve was advanced (Fig. 4.9, protocol D); and (C) shows the two-pulse response curve in which the whole response curve was delayed by the first pulse (Fig. 4.9, protocol E). The horizontal arrow in (A) shows the amount of advance of the PRCs obtained in (B) and the amount of delay obtained in (D) [56].

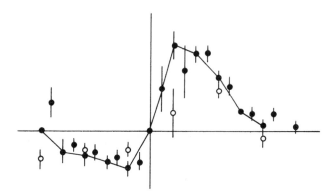

Figure 4.11 The phase responses due to a second pulse (open unconnected circles are the delay data from (C) of Fig. 4.10; filled unconnected circles are the advance data from B of Fig. 4.10) fall on the same curve as the single pulse curve (filled circles connected by lines; data from A of Fig. 4.10) when the phase shift due to the first pulse is subtracted from the phase shift measured after two pulses [56].

EXPLANATIONS BASED ON PRCS

One of the reasons that phase response curves captured the interest of circadian biologists is that they offered elegant explanations of fundamental observations of the properties of circadian rhythms, such as the effects of constant conditions on period length and the mechanism of entrainment.

First, PRCs have been used to explain the effects of light intensity on period length and the alpha/rho ratio. In Chapter 2, data were presented illustrating the circadian rule—the fact that in constant light, diurnal organisms (such as sparrows) have a shorter period length than they do in constant dark and that nocturnal organisms (such as hamsters) have a longer period in constant light than they do in constant dark. Many nocturnal animals (e.g., flying squirrels, hamsters, mice) have a larger delaying portion than advancing portion of their PRCs. Thus, in the absence of a Zeitgeber, the net effect of light should be delaying, producing a longer period length in constant light. Diurnal animals (e.g., house sparrows) have a larger advancing portion than delaying portion of their PRCs. Thus, in the absence of a Zeitgeber, the net effect of light should be advancing, producing a shorter period length in constant light [119, 256].

Second, Pittendrigh [258] offered an "interpretation of entrainment in terms of response curves for light."

> "The essential feature of this interpretation is that the entrained steady-state represents an equilibrium between the tau_{DD} and the action of light which can be either an advancing or a delaying effect. It follows that in a typical diurnal species (like a bird or lizard) . . . an equilibrium will also develop if . . . at equilibrium the advance section of the response curve (lies) within the light period to offset the daily delay due to tau_{DD} . . . as photoperiod changes . . . its expansion in the spring forces the nocturnal species to the right following sunset, because the photoperiod now embraces more of the delay curve conversely it appropriately forces the diurnal species to the left (following dawn) because more of the advance curve is covered." [256]

These ideas constitute the "theory of entrainment by photoperiod" [256, 260]. The emphasis in the theory of entrainment by photoperiod is upon the role of light as it impinges on phase response curves for light pulses.

I have suggested that in entraining to changing daily photoperiod, an organism is not limited to using its phase shifting response curve that is sensitive to light. Thus I suggest a *theory of entrainment by photoperiod and scotoperiod*. In particular, my view explains entrainment when the days are shortening by invoking sensitivity to dark. Where studied, there has been found a response curve of sensitivity to dark pulses as well as to light pulses, and that the response curves have roughly opposite shapes [49, 200].

> "At most northern latitudes the days lengthen from the winter solstice to the summer solstice; the total length of light and dark remains 24.0h but the dawn

occurs earlier and the dusk occurs later. We suggest that the additional light acts on the circadian clocks in the manner prescribed by the LP-PRC to synchronize to 24h (e.g., net advances of the sparrow's innate longer-than-24h clock produce entrainment to 24h). In addition we suggest that light advances the activity onset and delays the end of activity to anticipate the next day's lengthened photoperiod. Conversely, from the summer solstice to the winter solstice the days shorten, with later dawn and earlier dusk. We suggest that the additional dark entrains the sparrows' clocks (net advances of the first innate longer-than-24h clock produce entrainment to 24h). Dark delays the activity onset and advances the end of activity so that the sparrows' pattern is altered to anticipate the next day's shortened photoperiod." [33]

This view emphasizes (1) the fact that both onset and end of activity are phase shifted to respond to the changing photoperiod; (2) that when days are shortening, dark sensitivity (as opposed to light sensitivity) is used; (3) the importance of the L/D and D/L transitions. It is easiest to visualize this working if one borrows terminology from one of the two-oscillator models (Chapter 8). Look at activity onset as controlled by one oscillator (a dawn, morning, or M oscillator), which is set by dark or light impinging on sensitivity curves, and the end of activity as controlled by a second oscillator (a dusk, evening, or E oscillator), which is set by light or dark impinging on sensitivity curves. The activity time, alpha, then becomes the time between the dawn and dusk setting (Fig. 4.12). The distribution of bimodal activity in response to photoperiod (Fig. 3.11) is of interest in this context. When photoperiod is very long (or LL) or short (or DD), the activity is more unimodal than at intermediate photoperiods where there is an increase in bimodality. Thus, if activity density represents oscillators, we could suppose that in extreme photoperiods the oscillators are forced into coincidence by the proximity of L/D and D/L, but that at intermediate photoperiods they have sufficient time separation to be expressed individually. Attractive as two oscillator models may be, however, it is still possible to imagine one-oscillator models (in which the shape of a single oscillation responds to photoperiod) that are consistent with the data (Chapter 8).

Third, PRCs have been satisfying to the extent that they can be used to explain the curious effects of skeleton photoperiods. The two phenomena that require explanation are (1) how an organism selects one of the two possible interpretations of a skeleton; and (2) how an organism chooses *dawn* and *dusk* in the bistability zone.

Regarding the selection of one of the two possible interpretations of a skeleton, as Pittendrigh points out, the choice is not ambiguous:

"Any skeleton is open to two interpretations: the cycle that defines an 8-hr skeleton photoperiod also defines one of 16 hr ... no matter which interval is seen first (8 hr or 16 hr), the phase relation ... of the pacemaker ... to the light cycle that develops is characteristic of the shorter (8 hr) photoperiod." [260]

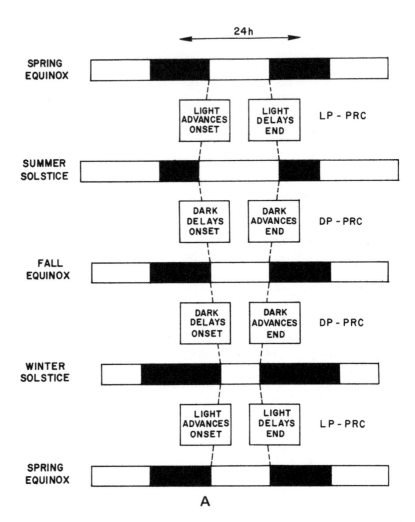

Figure 4.12 Figure 4.12A shows a rationale for the opposing appearances of the light- and dark-pulse phase responses offered by the author. The author has proposed that the opposing response curves permit an organism to reset both when the days are lengthening (June to December in the northern hemisphere) using the sensitivity revealed by the light-pulse response curve) or shortening (December to June) using the sensitivity revealed by the dark-pulse response curve [49, 200].

Figure 4.12B shows the same idea but illustrated with sparrows' actual light- and dark-pulse phase response curves to show the effects that light achieves on the onset and end of activity using both light and dark sensitivities (as mapped by PRCs).

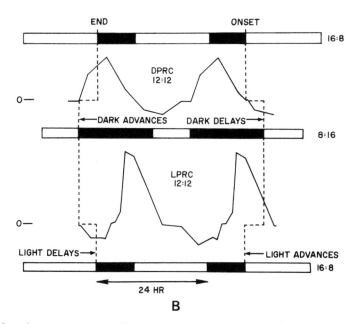

B

That is, for the sparrow to choose the 16h *photoperiod* from the skeleton, it would experience a light pulse falling about 8h into its subjective dark-time which, from the PRCs, would normally require a big advance rather than onset of activity at the expected chosen dawn pulse. Thus, the PRC suggests that the choice of a 16h photoperiod for the sparrow would result in an unstable situation and thus be precluded, whereas the alternate choice of an 8h photoperiod would produce stable entrainment and thus is the photoperiod of choice.

In the matter of assignment of dawn and dusk to pulses in the bistability zone (e.g., in LD1:11), when we investigated the sparrows' choices in DL1:11, [53] we found that their choice could be predicted by where the first pulse they experienced fell with respect to their PRC. If the first pulse fell in the early subjective night (delay portion of light pulse PRC), the sparrows delayed and began activity on the second pulse, which fell in the early subjective day. When the first pulse fell in the late subjective night, the sparrows advanced and began activity on the first pulse. A similar result was obtained by Pittendrigh and Ottesen using *Drosophila* eclosion [260].

RESETTING GENERALITIES

By comparing the properties of PRCs collected from a variety of species using a plethora of signals, we can arrive at some generalizations about the characteristics of PRCs (Fig. 4.13), a kind of PRC *atlas* [3]. I offer the following list of generalizations about the responses to light and dark pulses:

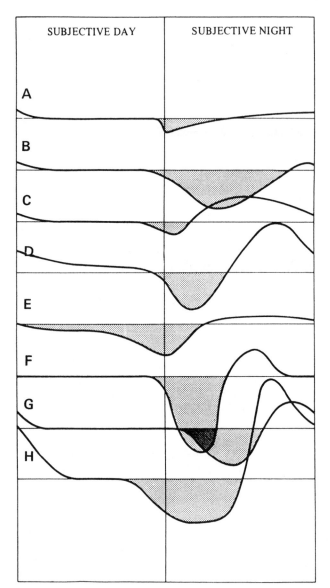

Figure 4.13 Brady [79] plotted phase response curves for a number of species showing the dead zone on the left in the subjective day, delays in the early subjective night usually encompassing the L/D, and advances in the late subjective night. (A) flying squirrel; (B) house mouse; (C) hamster; (D) cockroach; (E) ground squirrel; (F) flesh fly; (G) fruit fly; (H) dinoflagellate (redrawn).

1. "The time at which the stimulus is given (i.e., when during the animal's sleep-wake cycle), determines whether any response occurs, and if so, whether it is an advance or a delay [148]."
2. The details of the responses depend upon the species, whether the organism is nocturnal or diurnal, the pretreatment lighting regimen (photoperiod, T), and the nature of the pulse signal.

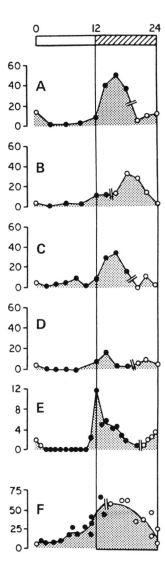

Figure 4.14 A way to represent phase shifts that shows *sensitivity* without concern for the direction of the shift (advance or delay) is to plot the absolute values for the shifts as was done here by the author [37]. (A, C, D) three species of mice; (B) hamster; (E) flying squirrel; (F) house sparrow. The author has further suggested that the sensitivity might correlate with some physiological parameter, such as pineal NAT activity.

3. In response to light pulses, delay phase shifts (− delta phi) occur in the early subjective night and may encompass the *dusk* (L/D) transition. Advance phase shifts (+ delta phi) occur in the late subjective night and may encompass the *dawn* (D/L) transition.

4. In response to dark pulses, advances occur in the early subjective night, and delays occur in the late subjective night—the PRC for dark pulses is out of phase (and may be the crude *mirror image*) of the PRC for light pulses.

5. There is a *dead zone* in the subjective day when the system is relatively *refractory* or insensitive to pulse signals.

6. There is a change from advances-to-delays for the dark pulse PRC or a change from delays-to-advances for the light pulse PRC that occurs more or less in the middle of the subjective night which may correspond with the *singularity* (Chapter 5).

7. The phase shift of the pacemakers caused by a resetting signal is near instantaneous (it is completed in less time than it takes the cycle in which it is given to complete its excursion).

As we pointed out in Chapter 3, light and dark are not the only Zeitgebers. For example, temperature is also a Zeitgeber for some rhythms in some organisms. Thus, it is possible to measure phase response curves for signals consisting of pulses of temperature, e.g., 20°C to 28°C for 4h, then back to 20°C [90]. Response curves for temperature pulses look very similar to those for light pulses with first delays and then advances and with dead zones.

The apparent discontinuity in phase response curves can be removed by plotting absolute values of phase shifts as I did (Fig. 4.14). I intended such plots as a measure of *sensitivity* to light; they are remarkably similar to other sensitivity plots (e.g., the pineal enzyme NAT) and clearly show the tendency for maximum phase shifting sensitivity to occur in the subjective dark-time.

5

CAN THE CLOCK
BE STOPPED?

"Martians . . . In three other points their physiology differed strangely from ours. Their organisms did not sleep, any more than the heart of man sleeps. Since they had no extensive muscular mechanism to recuperate, that periodical extinction was unknown to them. They had little or no sense of fatigue, it would seem. On earth they could never have moved without effort, yet even to the last they kept in action. In twenty-four hours they did twenty-four hours of work, as even on earth is perhaps the case with ants."

H. G. Wells [375]

"Much is known about the 24-hour clock in lower animals—rats, hamsters, and monkeys—enough to predict with some degree of assurance the sequences and interrelations of events before and after the discovery of fire by man, and to bring out the great effects that discovery of fire must have had on intellectual and cultural evolution of *Homo erectus.* Observations obtained in this way show that the effect produced by discovery of fire on man's 24-hour clock could be the most far-reaching single cultural event in his evolution [462]."

Curt P. Richter [284]

DISRUPTING THE CLOCK

In this chapter we discuss the possibility of and the consequences of disrupting the operations of a biological clock. We know that rhythmic processes can be stopped and started. For example, if you take a beating turtle heart preparation and place it in the refrigerator, the heart will slow down and stop beating. Remove the heart from the refrigerator a day later and as it rewarms it commences its beating. This process can be repeated several times. In this chapter disruptions do not include alterations of freerunning period and resetting (discussed in the previous chapters). The disruptions discussed here do include apparent *stopping* of the clock, *arrhythmia*, and *splitting* of rhythms into multiple components. The chapter covers the variety of means of disrupting rhythms—lighting, temperature, surgery, chemicals. Since light is the main Zeitgeber we have been discussing for circadian rhythms, it is appropriate to begin with the disruptive effects of light on circadian rhythms. The disruption of circadian rhythms by light may be of more than academic interest because ever since the discovery of fire we have exposed ourselves to non-natural lighting regimens, and artificial lights at night are a fact of life for us and for the animals that are our neighbors (Figs. 5.1, 5.2). I have lumped together discussion of some phenomena that I viewed collectively as disruptions of biological clocks. Placing these disruptions together in this chapter is not meant to imply that they have a common cause, nor that there are not other phenomena that might be considered disruptions.

Figure 5.1 The use of fire made it possible for man to add light to his environment.

Figure 5.2 City lights provide an artificial light extension (increasing the photoperiod).

CONSTANT LIGHT, ARRHYTHMIA

Some data that made a big impression on me when I first entered this field were the response of house sparrows to constant light. The perching activity of sparrows placed in constant bright light (e.g., 1000 lux) becomes arrhythmic—the sparrows hop with gusto all 24h day after day, month after month, with no obvious ill effects (Fig. 5.3). In LL, the sparrows' total daily activity was not necessarily greater and the absence of rhythms was confirmed with computer analyses [51]. The ability of sparrows to be active throughout 24h for long periods of time raises many questions pertinent to humans, mainly whether it would be possible to increase our usable daily time and reduce sleep need by exposing ourselves to constant bright light. We may have already answered that question to some extent since, by use of artificial lamplight in the evening, we have extended our day to an average 16h, even during winter months when the days are short.

What is the meaning of arrhythmia in LL in sparrows? Does it mean that the clock is stopped? Or is the clock running but its expression is masked by a direct effect of light? I cannot answer that question now, and we may never be able to separate direct and circadian effects of light in sparrows. However, one group claimed that though blind sparrows respond to LL with shortened tau and lengthened alpha, it was not possible to obtain the arrhythmic response to LL in blinded sparrows even if very bright lights were used [224]. If that is the case, if follows that arrhythmia is due to masking and that masking effects of light in sparrows are perceived by the eyes. However, there is other evidence (mainly the phase setting effects of LL/DD transitions, Chapter 5) that can be interpreted to support the hypothesis that constant light does affect the pacemaking system.

We should not jump to the conclusion that living in extended lighting is

Figure 5.3 Very bright constant light causes arrhythmia. Here, 20 days of a sparrow's perching activity is shown; the sparrow was kept in constant light (LL, 870 lux). The sparrow was apparently arrhythmic; computer analyses of such apparently arrhythmic perching (periodogram, autocorrelation, power spectral analysis) did not reveal hidden rhythms in such data [30]. Each line of data is 24h long. When sparrows are placed in constant dark after being arrhythmic in constant light, a line through the onsets of their freeruns extrapolates to time points 12–16h after the time of lights-out (L/D). The author has suggested that constant light stops the sparrows' clocks at the end of subjective day so that, when the sparrows are placed in the dark, the circadian clock starts, they experience a subjective night, and the next subjective day begins 12–16h later [36, 59].

necessarily desirable and healthy either for us or for the animals and plants that are exposed to our artificial lights. Especially for plants, damage due to LL has been observed—e.g., failure of flowers to open in the evening primrose, fungi unable to form sporangia, loss of chlorophyll in an alga [90]. Interestingly, some of the deleterious effects of LL on plants can be prevented by exposing the plants to a daily temperature cycle while they are in LL.

Often, upon placing an organism in LL, the circadian rhythm is not lost immediately. Instead, the rhythm damps out more or less gradually over a few cycles, a so-called *fade-out* [90]. However, it is not always the case that LL produces arrhythmia, even when the LL produces damping of the oscillation.

CONSTANT LIGHT, SPLITTING

Not all organisms become arrhythmic in constant light. When hamsters are placed in constant light, as we have said their period lengthens and their activity time, alpha, shortens or is *compressed*. When male hamsters were kept

Figure 5.4 Some animals' circadian rhythms can *split* into two components. The record that illustrates this is for a hamster kept in LL—splitting occurred about 92 days into the regimen. Each line of the record is 48h of data. The investigators called one component of the split rhythm the morning oscillator (M) and the other component the evening oscillator (E). (Redrawn from Earnest and Turek [127].)

in constant light for a long time (e.g., 30–100 days), 56 percent of their circadian rhythms *dissociated* or *split* into two freerunning components (Fig. 5.4). The two components have different periods during the few days they take to separate, then they run in parallel 180 degrees apart (we note that this is unlike the data for two birds where two freerunning periods were obtained; see Chapter 2). Splitting was reversed within one to four days when the hamsters were returned to DD, the activity rhythm *re-fused* into a single component, usually with a longer tau than for the dissociated rhythms [127]. The scientists who have studied splitting have attributed it to the presence of two oscillators, a morning (M) and evening (E) oscillator. Splitting has been observed in species other than hamsters, such as tree shrews and rats.

Whether there is any evidence for split rhythms that relates them to bimodal patterns remains to be seen. The bimodal patterns differ from the split rhythms primarily in the sense that activity is seen in the interval between the two peaks (Chapters 1 and 3).

CONSTANT LIGHT, RESETTING

As I said, when trying to explain arrhythmia, it might be impossible to separate clock stopping from disruption of the clock due to masking. One question we can ask is: "what time do the organisms think it is in LL?" We can ask this experimentally by transferring animals from LL to DD and determining the phase of the onset of the freerunning activity in DD. When we did this with sparrows, we found that first the sparrows began a rest period at the L/D. About 12h later, some of the sparrows commenced a circadian rhythm in perch-hopping activity, which continued to freerun from that phase in DD [37]. The average duration of the rest period before the first perching was 15h [53]. The data are evidence that the circadian clock in sparrows is stopped at the end of the birds' day, the time the birds expect dusk to occur. However, there is an alternate, more complicated explanation—to wit, the biological clocks of arrhythmic sparrows are present and at different phases, but are masked, and the L/D transition resets the individual sparrows by different amounts so that, after L/D, they all appear to have the same time setting.

We applied the logic of placing sparrows in DD to assess their clock status to sparrows in LD1.5:1.5—would the sparrows' rhythms proceed as though begun at the last D/L, 1.5h before they were placed in DD? The answer was similar to that for LL; the sparrows' clock-time zero was 15–16h after the last L/D (Fig. 5.5) [53].

Sparrows are not the only organisms reset by L/D after LL. The effect is so reliable and routine that it is used to synchronize the rhythms of flies within a population of *Drosophila* [256]. How long does the LL have to be? Pittendrigh exposed *Drosophila* to LD12:12, then a final photoperiod of varying duration from 1–33h, then to DD to assess the phase of their freerunning rhythms (Fig. 5.6).

Using any final photoperiod longer than 12h, the flies' eclosion rhythm was timed by the L/D (the new phase was always 15h after L/D or 15h plus some multiple of 24h). So for flies, 12h of light was sufficient for L/D to set the clock, the L/D transition was an "absolute phase-giver" [256]. If the clock is stopped by light, then similar logic applied to long photoperiod means that in long photoperiod (e.g., more than 12h of light for *Drosophila*) the clock reaches a time point where it stops and that time point is similar to the stopping point in LL.

Chandrashekaran, Johnsson, and Engelmann [99] did an experiment with fruit flies to further examine the roles of *dawn* (D/L) and *dusk* (L/D) in the way that light pulses reset circadian rhythms. They found that the eclosion rhythm of the flies was set by L/D in the first half of the subjective night and was set by D/L in the last half of the subjective night (Fig. 5.7).

A SINGULARITY

Winfree [383, 384, 386] and other investigators have methodically measured phase response curves (Chapter 4) varying the signal duration of the pulse (e.g., from 0–120 sec). Plotting the data in three dimensions produces a *resetting surface* with a corkscrew shape. Winfree points out that there is:

> "...a curious implication of the corkscrew shape of the resetting surface. A corkscrew surface has a singularity, a central axis along which the slope is

Figure 5.5 Very short cycles may produce phase setting effects similar to constant light. The record is perching activity of a sparrow kept in LD1.5 : 1.5 in which it was active in the 90 min light periods (A) and did not display a circadian rhythm. When the sparrow was placed in constant dark at (B), its activity rhythm began to freerun with the onset of activity (filled circle) 11.5h after the last L/D (open circle) [53].

infinite. The theta-contour lines of the corkscrew surface converge on this central axis, which apparently represents a critical stimulus time, $T^* = 6.8$ hr, and duration, $S^* = 50$ sec . . . Taken literally, this feature of the resetting surface seems to mean that there is an isolated perturbation following which there is either no circadian rhythm of emergence in the steady-state, or one of unpredictable phase." [100, 383]

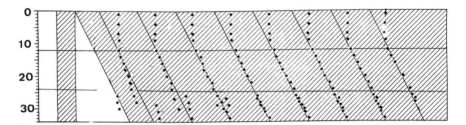

Figure 5.6 The dots show the phases of eclosion in populations of fruit flies with respect to the final photoperiod they experienced (wedge-shaped area). The numbers on the vertical left represent hours of final photoperiod. The horizontal axis is time with diagonal lines drawn at 24h intervals beginning 16h after the time of the last L/D when the first emergences occur. The experiment shows that the L/D determines the phase after a single photoperiod 12h or longer. Transients can be seen in the first three cycles after the L/D. (After Pittendrigh [256, 259].)

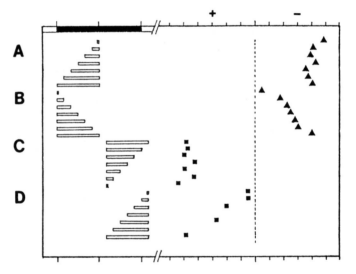

Figure 5.7 An experiment with fruit fly eclosion shows that the flies follow the L/D of a pulse during the first half of the subjective night and the D/L of a pulse during the last half of the subjective night. (After Chandrashekaran, Johnsson, and Engelmann [99].)

More recently, Enright and Winfree redefined *phase singularity* as "a point at which phase is ambiguous, and near which phase can take on any value" [149]. To place the time in context, in Winfree's protocol, 6.8h is measured from LL/DD, placing it in the middle of the night (sparrow c.t. 18–19h) in the region where many organisms' phase response curves change from advances to delays.

When *Drosophila* populations were exposed to pulses near the singularity pulse, a curious phenomenon occurred. The flies emerged, the eclosion rhythm was lost or *annihilated*. There are two alternate explanations for the result. (1) One possible explanation is that slight differences in the clocks of the individual flies that make up the population account for their variable responses to T^*S^*—that is, the stimulus hits some flies still in the advancing portion of their response curve while hitting other flies in their delaying portion—so that the flies are scattered to different individual phases. In this explanation, individual flies have not lost their rhythms, they have just been desynchronized one from another, and the singularity is just a trivial result of the fact that the experiment is done with a population of animals. (2) A second possible explanation is that emergence rhythms were restored after T^*S^* with a 120 sec light pulse 6, 12, 18, or 24h later; the peaks of these eclosions all fell at the same time. This is reminiscent of the result when sparrows were transferred from LL (where they were arrhythmic) to DD, and might mean that the singularity and LL both stop the clock. In this second view, T^*S^* drives the clock to a *phaseless* state where it remains until the second pulse reinitiates the oscillation and it can assume any phase.

DISSOCIATION

Arrhythmia is not the only type of disruption that has been reported for circadian clocks. There are data in the literature in which several components of a circadian rhythm have drifted out of phase with one another or several rhythms within an organism have been found to freerun with separate period lengths. The classic example of this is dissociation of the body temperature rhythm from the sleep-wake rhythm that has been reported for some freerunning humans (Fig. 10.7). Dissociation differs from splitting reported for hamsters in that the period lengths of the components are not the same.

The explanation of dissociation is that the organism contains multiple circadian oscillators. Normally, these rhythms have the same period derived either one from another or from a common pacemaker. The word *coupling* refers to the physiological relationship by which two oscillations within an individual are synchronized.

Dissociation has been claimed for freeruns but also dissociation has been claimed as occurring during phase shifting (e.g., in jet lag). The idea is that separate rhythms resynchronize at different rates depending upon the strength of their coupling to the pacemaking rhythm. An explanation of the physical problems that are observed when rhythms are disrupted by travel and shift work (fatigue, accidents, headaches, burning eyes, gastrointestinal problems, shortness of breath, sweating, nightmares, insomnia, menstrual irregularities [215]) is that they result from asynchrony among internal rhythms that would normally be synchronized.

TEMPERATURE

The effects of temperature upon biological processes has been a subject of interest for the fields of hibernation, cryobiology, thermoregulation, and biochemistry. There are also some disparate aspects of temperature that pertain to circadian rhythms. There are circadian rhythms of body temperature (Chapter 1), temperature can act as a Zeitgeber (Chapter 3), cold affects the expression of rhythms, and rhythms are relevant to hibernation. First, I will discuss some aspects of temperature in biology, then I will consider some specific ways in which temperature has to do with biological rhythms.

The machinery of biological organisms is dependent on chemical reactions, and the effects of temperature on reaction rate are well established [195]. In general, heating increases the rate of a reaction, which would provide a rationale for the evolution of organisms' ability to generate heat and regulate their temperatures above the temperatures in their environment. However, scientists argue that biological reactions would still be too slow at normal temperatures for cells to function, so there are other biochemical strategies for increasing reaction rates such as concentrating the reactants and the use of catalysts. Living things have catalysts called *enzymes* that they use to increase reaction rates. Typically, the activity of an enzyme is negligible near freezing (Fig. 5.8) and increases with increasing temperature (about double for each 10°C increase). Scientists quantitate the dependency of reaction rate on temperature with a value, Q_{10}, that represents the effect on the rate of a reaction of a 10°C temperature increase. If the reaction rate doubles, $Q_{10} = 2$. If there is no effect of temperature on a reaction, $Q_{10} = 1$. At high temperatures (e.g., 40–60°C), the enzymes break down (are inactivated, a process called *thermal denaturation*). We routinely use cold to arrest the rate of biological reactions. For example, the activity of the pineal enzyme, N-acetyltransferase, can be preserved by freezing pineal glands or by maintaining homogenates of pineal glands on ice [44].

Figure 5.8 Many biological processes stop when an organism is placed in the cold or is frozen.

What are some of the effects of temperature on circadian rhythms? First, as might be expected, extreme heat or cold abolishes rhythmicity (Fig. 5.9A, G). Second, amplitude is altered (Fig. 5.9). However, there is surprisingly little effect of temperature upon period length (Fig. 5.9, Tables 5.1, 5.2) [232]. Q_{10} for circadian rhythm period length is usually near 1 (0.9–1.2). At first this evidence appears to indicate that the circadian period length is *independent of temperature* and not dependent upon a chemical reaction; rather, it implicates a physical process such as diffusion. However, because the Q_{10} is not exactly 1, rhythmologists have argued for a process of *temperature compensation* in which two or more opposing processes are involved so that the net effect is very little effect of temperature [90]. Pittendrigh has argued for the adaptive value of a temperature compensated clock:

> "The facts that the values are not exactly 1.0, and that some are less than 1.0, clearly imply that the effect is achieved by some compensatory mechanism rather than by control through a limiting factor that is absolutely temperature independent. This temperature compensation of the living oscillator's frequency is clearly an essential prerequisite for effective entrainment; had it the usual Q_{10} near 2.0, it would commonly fall outside the limits within which the light cycle could hold it, by entrainment, to the frequency of the environmental cycle. Temperature compensation is also an essential feature in any oscillation that is to be exploited as a useful time-measuring system." [257]

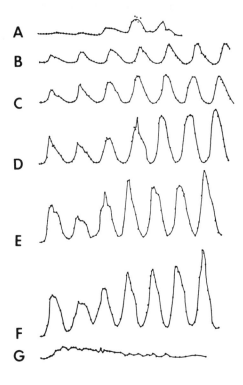

Figure 5.9 Circadian rhythms respond to temperature. The main response is in the presence of a rhythm and amplitude of the rhythm. Amazingly, the period length exhibits only small changes in response to temperature. The figure shows graphs of bioluminescence in a culture of dinoflagellates. The horizontal axis is 180h; the vertical axis measures luminescence; the cultures were in constant light. A = 32°C, B = 26.8°C, C = 23.6°C, D = 22°C, E = 19°C, F = 16.5°C, G = 11.5°C. The rhythm damped at very warm temperatures (e.g., 32°C) and was virtually lost in the cold (11.5°C). (After Sweeney and Hastings [325, 326].)

TABLE 5.1   Period Length is Only Slightly Altered by Temperature[1]

Species	Temperature Degrees C	Period length (h)
Periplaneta	19	24.4
	29	25.8
Gonyaulax	16	22.5
	22	25.3
	32	25.5
Phaseolus	15	28.3
	20	28.0
	25	28.0
Lacerta	16	25.2
	25	24.3
	35	24.2
Neurospora	24	22.0
	31	21.7

[1]After data collected from various sources for cockroach activity, bean leaf movements, bioluminescence, lizard activity, and bread mold growth [90, 326].

TABLE 5.2 Q_{10} Values for Circadian Period Length[1]

Q_{10}	Organism	Rhythm
0.8	alga	sporulation
1.0	bats	activity
1.0–1.3	beans	leaf movement
1.0	bees	feeding time sense
1.0	bread mold	growth zonation
1.0	ciliate	mating
1.0	crayfish	eye-pigment migration
1.0	fiddler crab	color change
1.0–1.1	flagellate	phototaxis
1.0–1.2	fruit flies	eclosion
3.0	grasshopper	hatching
1.1	hamster	activity
0.9	marine dinoflagellate	bioluminescence
0.9	marine dinoflagellate	cell division
1.1–1.4	mouse	activity
1.3–1.5	mold	sporulation
1.0	oats	growth rate
1.0	roaches	activity
1.1	sunflowers	exudation

[1]After values in Sweeney and Hastings [326].

Since environmental temperatures fluctuate greatly with weather changes, it seems logical that for any clock to be precise, its rate must be relatively independent of temperature. Bünning [90] has pointed out, however, that in tropical regions with relatively constant temperature, the adaptive value of temperature compensation is less important than at other latitudes and points to *Phaseolus mungo* as an example of a tropical plant whose period lengthens beyond 32h at 17°C.

I placed this section on temperature in the chapter about *clock stopping* for a reason. Organisms, especially those that are not warm-blooded such as the invertebrates, can be cooled to temperatures just above freezing (e.g., 0–5°C). Such cold temperatures retard (delay) the clock for more or less the length of time the cold endures (e.g., for spiders, roaches, beans); some organisms' clocks are very resistant to cold temperatures (e.g., alga *Oedogonium* and pond skater *Velia*). After prolonged cold, rewarming usually has a phase setting effect (e.g., in roaches [288] and bean plants [90]). Because the phase shifts were not always the length of the chilling, the investigators have argued that the clock was not stopped.

Warm-blooded animals (the birds and mammals, *homeotherms*) maintain body temperatures in the vicinity of 37°C. There are circadian fluctuations in body temperature (about 1°C in a human, as much as 5°C in a sparrow). Birds have higher body temperatures than mammals. Some homeotherms are able to drop their body temperatures on a nightly basis (*daily torpor*) or a seasonal basis (*hibernation*). It is possible to override normal thermoregulation and to artificially chill mammals in the laboratory by anesthetizing them (with ether or sodium pentobarbital) and placing them in a pan in an ice bath. Rawson [274] did this with mice and hamsters and found that chilling the animals to as low as 14°C for 1–8h delayed their subsequent activity rhythm up to 124 min. With these experiments and others using bats, the Q_{10} for mammals was estimated as near 1 as for other organisms.

Naturally, as I mentioned, some organisms are able to lower their body temperatures on a daily or seasonal basis. For example, hibernators reduce body temperature in *bouts* (lasting a number of days each) throughout the winter season presumably as an energy saving mechanism to permit them to survive winter weather and food shortage extremes. During the season of hibernation, the animals reduce body temperature so that it is a few degrees above the ambient temperature in their burrows (*hibernacula*). A number of investigators have examined hibernating bats and ground squirrels and found, as Bünning summarizes, that "Warm-blooded animals . . . apparently show an errorless continuation of the clock, even during hibernation, when temperature is lowered. They tend to awake at a time of day characteristic of their normal activity phase" [90].

In summary then, while the circadian clock responds to temperature as a Zeitgeber, and the amplitude of rhythms is subject to modification by temperature, the period of the clock is remarkably resistant to changes in environmental temperature.

ABLATION

There are a number of surgical procedures performed upon organisms that have produced *arrhythmia*, a loss of circadian rhythms, in constant conditions. The investigators who have done these procedures have argued that the arrhythmia that is produced may represent the removal of the circadian pacemaker, or, at the very least, disconnection the measurable rhythm (the *hands*) from the underlying pacemaking oscillation (the *clock*). In these cases, the argument is that the clock is not stopped but has been removed. The possible biological clocks discovered in this manner are discussed with their respective organisms in Chapter 9.

CHEMICALS

Arrhythmia, or loss of circadian rhythms, has also been produced in organisms with various chemical (or pharmacological) treatments. For example, the circadian locomotor rhythm in house sparrows is readily abolished by providing a continuous supply of melatonin in the drinking water or delivered by a constant release capsule implanted in the bird (Fig. 5.10) [54, 347]. While the amplitude of many observable rhythms can be reduced with chemical inhibitors (e.g., sodium cyanide, arsenate, 2,4-dinitrophenol, sodium fluoride) [90], circadian period length has been remarkably resistant to temperature, so much so that a kind of *biochemical compensation* has been proposed [263]. Similar to the argument for temperature compensation, clock precision requires relative stability to maintain timing in the presence of environmental fluctuations. A typical discussion of the question of clock stopping by chemicals appears in Bünning's discourse on the "Effects of extreme reduction of respiration":

> "The clock stops only when respiration has dropped to a very low percentage of the intensity under optimum temperatures. We must draw the same conclusion from experiments on the influence of respiration inhibitors. On the other hand, an inhibitor restricting respiration may prevent the overt periodicity without delaying the clock itself. One should not conclude from this that the operation of the clock is completely independent of respiration or that it depends only upon a certain component not being affected by the inhibitor. Yet it is safe to say that the amount of energy required for the continuation of the clock is smaller than that for carrying out the controlled processes (e.g., growth, running activity, and spore discharge). The inhibitor concentration must be close to lethal to stop the clock by reducing respiration. If this degree of inhibition is accomplished, one can in fact observe effects upon the course of the clock." [90]

I have listed some effects of chemicals on circadian rhythms in Table 5.3.

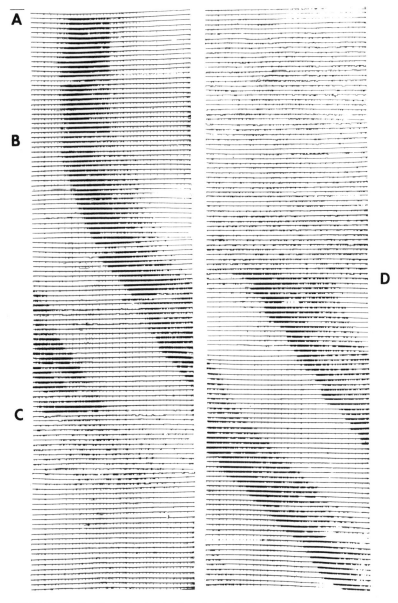

Figure 5.10 Continuous melatonin administration produces apparent arrhythmia in sparrows. The perching activity of a sparrow is shown; each line is 24h of data; the record continues in the second column. In (A), the sparrow was kept in LD6 : 18, lights-on 6 a.m. At (B), it was placed in DD and freeran with a period greater than 24h. At (C), the sparrow was implanted with a Silastic capsule which constantly released melatonin—the sparrow appeared to become arrhythmic and did so immediately. When the capsule was removed at (D), the sparrow resumed its freerun. Similar results can be obtained by providing melatonin in the drinking water [54, 347].

TABLE 5.3   Chemical Alterations of Circadian Rhythms[1]

Chemical	Alteration	Rhythm
arsenate	suppression[2]	sporulation, *Oedogonium*
CO_2 increase	inhibition[3]	bean leaf movement
colchicine	transient long tau	beans
cycloheximide	increase tau	phototaxis, *Euglena*
D_2O, heavy water	lengthens tau	*Euglena, Phaseolus,* mammals
dichlorophenyl dimethyl urea	suppression	photosynthesis, *Gonyaulax*
2,4-dinitrophenol	suppression	sporulation, *Oedogonium*
ethyl alcohol	lengthens tau	beans, *Phaseolus*
ethyl alcohol	lengthens tau	isopod, *Excirolana*
melatonin	arrhythmia	activity, *Passer*
NaCN	suppression	sporulation, *Oedogonium*
NaF	suppression	sporulation, *Oedogonium*
oxygen removal	suppression	bean leaf movement
O_2 replaced with N_2	transient delay	oat coleoptile growth, *Avena*
puromycin	inhibition	luminescence, *Gonyaulax*
testosterone	lengthens tau	activity, mouse
theobromine	lengthens tau	beans, *Phaseolus*
theophylline	lengthens tau	beans, *Phaseolus*
urethane	transient long tau	beans

[1]References [90,113,146]
[2]Suppression means reduction of amplitude
[3]Inhibition means blocks rhythm

CAN THE CLOCK BE STOPPED?

The title of this chapter posed a question as to whether it was possible to stop a circadian clock. The question was posed as a scientific question for which experimental evidence can be sought. The reader can accuse me of asking a rhetorical question (asked to emphasize a point or introduce a topic, that is, with no answer truly expected) because I have been unable to produce an unequivocal answer. In the chapter I reviewed the effects of various treatments (light, temperature, surgery, chemicals) that modify rhythms (reduce their amplitude, change their period, perhaps abolish them). A main point made was that while the clock is readily *reset* by light, its period length is surprisingly impervious to temperature and chemical treatments. The most evidence for clock stopping comes from studies using constant light or critical (singularity) pulses, but proof that these treatments stop the circadian pacemaker is, as we have seen, elusive. There are alternative explanations for apparent clock stopping (rhythms are masked, the measured rhythm is uncoupled from the clock, rhythms are dissociated, etc.) that are difficult to rule out.

6

CALENDAR

"It would, therefore, appear that . . . daily increases of illumination are conducive to developmental changes in the sexual organs."

William Rowan [299]

"Analysis of the photoperiodic phenomena has shown that it is not the length of the day as such that is decisive. The effects under consideration . . . can also be obtained by giving short days instead of long days and dividing the nights into two dark periods by a relatively short supplemental light (e.g., one hour or much less) . . . Other evidence has also led to the conclusion that it is not the length of the light period . . . that is decisive, but the length of the night. One possible interpretation of these findings is that a time-measuring process is initiated at the beginning of the light period or at the beginning of the dark period. At a certain number of hours after the beginning of light or darkness, this process induces a sensitive stage that responds specifically to light . . . There is ample proof that the strongest effect is not obtained by simply dividing the dark period into two equal parts with such a light interruption. Rather, the time of highest sensitivity has a distinct relation either to the beginning of the light period or to the beginning of the dark period. This may be asymmetric, depending upon the length of the dark (or light) period."

Erwin Bünning [90]

PHOTOPERIODISM

Most of the book so far has dealt with circadian biological clocks. Many organisms also have a biological *calendar* that they can use to time seasonal events. We enjoy the spring progression of flowers (first the snowdrops, then the crocuses, next the daffodils, followed by the tulips, etc., Fig. 6.1), and most of us are aware that other events as well correlate with changes in season, especially at the temperate and arctic latitudes. Temperature can provide seasonal information, but that information is less precise than the information provided by the daily changes in the times of dawn and dusk. For example [234], at 60 degrees north latitude, the longest day of the year (*summer solstice*, the first day of summer) is about 18h and the shortest day of the year is about 6h (*winter solstice*, the first day of winter). But at 30 degrees, daylength at the winter solstice is only about 10h and at the summer solstice is about 14h. Near the equator the light cycle is close to LD12:12 throughout the year, and in the arctic regions extremely long days and nights occur. At tropical latitudes in the vicinity of the equator, seasonal changes are minimal and possession of a calendar may be less important to survival than in temperate and polar regions.

There is scant evidence that humans possess a biological calendar, but even prehistoric human cultures may have gone to great lengths to make calendrical calculations (Fig. 6.2) and may have erected monuments like Stonehenge for astronomical calculations. It is easy to make a logical argument that calendar information had predictive value for primitive agriculture and animal husbandry. The home gardener makes routine use of planting and flowering schedules published by the nurseries in planning his landscaping.

As I discussed, dawn and dusk provide cues as to time of day that can be used to reset circadian rhythms. At most latitudes, day and night length

Figure 6.1 Many biological events are associated with particular times of year, such as the blooming of daffodils in springtime.

change precisely and systematically throughout the year (Table 6.1). Thus, the changing times of dawn and dusk—which methodically alter daylength (*photoperiod*) and nightlength (*scotoperiod*)—contain seasonal information as well as time of day information. Organisms can use this information as a calendar to time seasonal events in their physiology. Palmer [250] defined photoperiod-

Figure 6.2   The complex and beautiful calendar is testimony to the importance human culture has placed on keeping track of annual events.

TABLE 6.1   Annual Changes in the Environment[1]

Month	Temperature[2]	Rain[3]	Snow[3]	Photoperiod[4]
January	26, 40	3.26	6.3	9h 35 min
February	26, 41	3.08	6.7	10h 49 min
March	33, 50	3.54	3.9	12h 8 min
April	43, 62	3.31	0.2	13h 35 min
May	53, 73	3.35	0.0	14h 46 min
June	63, 81	3.64	0.0	15h 17 min
July	68, 85	4.11	0.0	14h 51 min
August	66, 83	4.51	0.0	13h 42 min
September	60, 77	3.39	0.0	12h 17 min
October	49, 66	2.82	0.1	10h 52 min
November	39, 54	3.10	0.7	9 h 38 min
December	29, 43	3.19	3.9	9h 5 min

[1]In the vicinity of Pennsylvania
[2]Minimum, maximum, degrees Fahrenheit
[3]Inches
[4]From the *Farmer's Almanac* for 20th of the month, 1988, in Boston [339]

ism as "A response, such as reproduction or migration, of an organism to the relative length of day and night." Table 6.2 lists a number of events that are photoperiodic.

The list in Table 6.2 and the experiments in the literature tend to reflect the fact that plant flowering, insect diapause induction, and animal reproduction are easily studied phenomena with clear biological markers. It is quite easy to make rationales for particular organisms' seasonal adaptations that permit them to anticipate and make best use of the advantages of spring and summer (e.g., to raise their young) and to avoid the harshness of winter.

It is possible to determine experimentally for any given photoperiodic event the point at which a short day (long night) is distinguished from a long day (short night). The phrases *critical day length* or *critical photoperiod* refer to this; most measured critical photoperiods fall between 10h and 14h of light. The critical photoperiod within a species varies systematically with latitude. For example, the critical photoperiod for diapause in a butterfly, *Acronycta rumicis*, is 15h at 43 degrees north latitude but 18h at 50 degrees north latitude.

In considering photoperiodism, there are several confounding factors. First, light has a *dual role:* (1) it defines the photoperiod, and (2) it sets the phase of circadian rhythms. As we shall see, interpretation of photoperiod by an organism is dependent upon the circadian rhythm, although some organisms seem to have separate physiological sites (e.g., photoreception in plant photoperiodism resides in the leaf blade but the leaf joints detect light for phase shifting the circadian rhythm [90]). Second, there is a problem with the routine jargon that tends to refer to photoperiod even in cases where it has been demonstrated that the darklength, or scotoperiod, is what is measured. A paragraph in Palmer points this out:

> "Short-day and long-day (blooming) plants were placed under short days. This of course, permitted flowering in the former and prohibited it in the latter. The dark portion of each cycle was then interrupted by a short interval of light. This simple treatment inhibited flowering in the short-day plant and permitted it in the long-day plant. This means that it is the length of the dark period that the plants are actually measuring, and so, disrupting it with a light break spoils the measurement. Therefore, the terminology that has become so thoroughly ingrained is incorrect; short-day plants are really long-night ones, and long-day plants really short-night plants." [250]

In this chapter I will discuss the so-called light-break experiments such as the one referred to above, and also show data for the way in which nightlength is measured. It is, however, traditional to use terminology that refers to light and daylength rather than night and nightlength.

TABLE 6.2 Photoperiodic Events [90, 250]

appearance of sexual plant lice	molting
bulb dormancy induction	reproductive cycles
bulb dormancy termination	seed germination
cambium activity	succulence
diapause in insects	testicular size
flowering	tissue differentiation
fur color change	tuber formation
migration of birds	vegetative development
migratory restlessness	

In many vertebrates, there are dramatic changes in the physiological events associated with reproduction that are associated with season and controlled by photoperiod. For example, the testis size changes dramatically—the testes of juncos are very small in the nonbreeding season (e.g., January) when the days are short, and grow dramatically when the juncos are placed experimentally in long days [299]. In my laboratory, golden hamsters kept in LD16:8 had testes weighing 3g a pair; but hamsters maintained in LD8:16 had testes weighing only 0.4g a pair (Fig. 6.3). The seasonal changes that occur in hamster testes have been diagrammed (Fig. 6.4). The critical photoperiod for hamsters is about 12.5h [142]. When the days shorten (or in the laboratory in short photoperiod, in the dark, or in blinded hamsters), the hamsters are *photosensitive* and the testes become small. The word *regression* refers to the process in which the testes not only show dramatic decreases in weight and

Figure 6.3 Many animals have seasonal cycles in their reproductive physiology. Hamster testes are small in the nonbreeding season (right) but enlarge when the hamsters enter their breeding season in March (left). The seasonal changes in hamster testes can be mimicked in the laboratory by keeping the hamsters on short photoperiod, which reduces the size of the testes to 0.22g each (right) or long photoperiod, which increases testis size to 1.45g apiece (left).

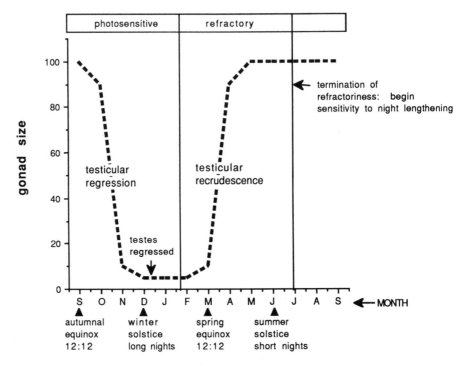

Figure 6.4 Illustration of the seasonal events that occur in the reproduction of an organism such as a hamster. The testes are large in September but regress (get smaller, stop producing sperm and hormones) in response to the shortening nights because the hamsters are "photosensitive." After a few months of short nights, the hamsters become "refractory" (insensitive to the effects of short nights) and the testes recrudesce (grow, become functional). In the summer, after exposure to long days, the hamsters again become sensitive to night lengthening. (After Stetson and Watson-Whitmyre [314].)

size, but also cease sperm production. The testes remain *regressed,* or involuted in a reproductively quiescent state in the winter (e.g., December through February). But as spring approaches, the testes begin to grow and become functional (*recrudesce*), even if the photoperiod stays short—the term *spontaneous recrudescence* refers to this. The time during which the hamsters' testes do not regress in response to short photoperiod (February through July) is called the *refractory* period. Long photoperiods maintain the size of the testes when the hamsters again become *photosensitive* (in July) until after the fall equinox (in September) when the short days again cause the testes to regress. Thus, from year to year, hamsters and other seasonal breeders oscillate between a reproductively active, breeding status and a reproductively quiescent condition. Photoperiod (or darklength) is a major factor in synchronizing the annual cycle with the seasonal changes in the environment. Seasonal breeding ensures that the young hamsters are born at a time of year propitious for their survival. Another example with which the reader may be familiar is the white-

Figure 6.5 White-tailed deer provide a good example of an organism with readily visible seasonal changes. The deer mate in the fall, fawns are born in spring, the fur molts to reddish brown in spring and to gray in fall, and antlers are shed in the fall and winter. (Courtesy of S. Binkley.)

tailed deer, which has seasonal cycles in breeding, fur molting, and antler production (Fig. 6.5).

LIGHT-BREAK AND RESONANCE EXPERIMENTS

Bünning (opening quotation) and others [37, 88, 90, 173, 250] have described hypotheses for photoperiodism that have 4 features.

1. There is an underlying circadian rhythm of sensitivity to lighting conditions.
2. Day (or night) length is determined by when light (or dark) falls with respect to the underlying circadian rhythm of sensitivity.
3. The duration of dark is the measured parameter.
4. Light (and dark) have a dual role in specifying photoperiod and in setting the circadian rhythm of sensitivity.

As suggested in Chapter 6 and the opening quotation, evidence for these ideas about how photoperiodism works comes from experiments with short pulses of light that have been used for both plants and animals. In this section we will discuss these experiments, dividing them into light-break (Fig. 6.6A) and resonance (Fig. 6.6B–E) experiments.

In a typical *light-break* experiment [88], groups of organisms are exposed to asymmetrical 24h skeleton photoperiods in which there is a long light pulse (e.g., 6h) and a short light pulse (e.g., 2h); the short light pulses in the different treatments are spaced at intervals scanning the 18h dark period. All of the

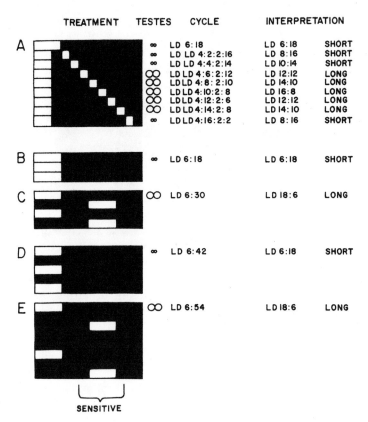

Figure 6.6 Circadian biologists have used protocols with light pulses to prove that the circadian clock is involved in photoperiodic time measurement.

(A) illustrates a *light-break* experiment in which a short light period (4h) is paired with a light pulse (2h); collectively, 2h of light is insufficient for the gonads (e.g., of a sparrow or hamster) to enlarge; however, as illustrated in the testes column, cycles in which the pulse falls in the middle of the dark period do result in large testes. This means that the timing of the light is important, not merely its duration. The duration of dark is important.

(B)–(E) illustrate a *resonance experiment* in which 6h light pulses are imposed in 24h cycles (B), 36h cycles (C), 48h cycles (D), and 60 cycles (E). The testes enlarge in the 36h and 60h cycles because the light falls in the organisms' subjective dark.

In other words, if light falls in the middle of the subjective night (sensitive period), the photoperiod is viewed by the organism as a long photoperiod, and the testes enlarge. The author interprets these experiments with the response of a pineal enzyme, N-acetyltransferase, whose light-inhibited night activity would reduce the production of a gonad-inhibiting hormone, melatonin during the sensitive time.

skeletons contain the same total light-time, 8h. The photoperiodic response of the organism (e.g., occurrence of diapause) exhibits a short-day response (100 percent diapause) when the 2h pulses are just after the 6h pulse and when they occur late in the dark period (just before the 6h pulse). But when the pulses occur in the middle of the dark (e.g., 14–16h after D/L of the 6h light), a long-day response (e.g., inhibition of diapause) is obtained. These experiments produce the same kind of results with diapause in insects, with flowering in plants, and with testes of hamsters. They clearly show that it is not just *how much* light (all cycles had 8h light) but *when* the light fell that was important to the organism's determination of whether to display its short- or long-day response.

Light-break experiments can be understood from what we know about skeleton photoperiods. In Chapter 3 I described entrainment to skeleton photoperiods and we noted that sparrows, for example, always interpreted LD1:6:1:16 as LD8:16, selecting the D/L as dawn that placed the second pulse 7h later. Here, we propose that the organisms entrain to the asymmetric skeletons in such a manner that the skeleton is viewed as a short photoperiod. However, a short photoperiod cannot be derived from the skeletons containing pulses in the middle of the night, so that the organisms respond as to a long photoperiod. The photoperiodic responses to the skeletons in a light-break experiment can be understood by replotting the data so that the D/L taken by the organisms as subjective dawn is aligned (Fig. 6.7). When this was done, evidence appears for the *length of the dark period* as the parameter that gives the photoperiodic time cue [228].

More evidence for circadian rhythm involvement in photoperiodism comes from another type of light pulse experiment, the *resonance* experiment (Fig. 6.6) [90, 143, 173]. In the resonance experiment, light pulses of shorter duration than the critical photoperiod (e.g., 6h) are imposed in cycles with varying durations that are or are not integer multiples of 24h (e.g., 24, 36, 48, 60h cycles in Fig. 6.6). The photoperiodic response of the organism (e.g., soybean flowering, hamster testis growth) is measured. In a resonance experiment, non-24h regimens such as 36h and 60h cycles produce the long-day response, whereas the cycles that are multiples of 24h (24h, 48h, 72h) produce the short-day response. As with the light-break experiments, the resonance experiments can be understood by considering when the light pulses fall with respect to the organism's circadian rhythm; as with the light-break experiments, those regimens that result in mid-night pulses produce a long-day response (Fig. 6.8).

PHYSIOLOGICAL BASES FOR PHOTOPERIODISM

The mechanisms underlying photoperiodism are no longer a total mystery. Good evidence points to the participation of particular structures in photoperiodic responses and to the means by which they measure photoperiod. It

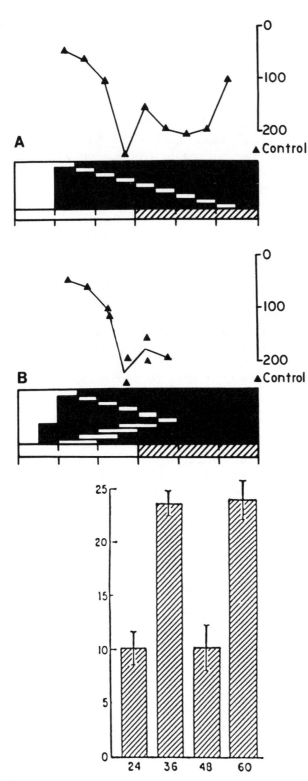

Figure 6.7 Data for sparrow testis response to skeleton photoperiods plotted to show a possible dependency upon length of the dark-time experience by the birds relative to their circadian rhythms. The lower graph is a replot of the data of the upper graph [36, 228].

Figure 6.8 Testis response of hamsters kept in the resonance experiment shown in Fig. 6.6 (B)–(E). The vertical axis is testis weight (mg/g) and the horizontal axis is cycle length [143].

seems likely that diverse organisms have capitalized on separate aspects of their physiology to produce the photoperiodic responses appropriate for their own individual environmental milieu. I have outlined above a photoperiodic mechanism in which the circadian clock operates, and I will develop the ideas further in the following sections.

However, there are photoperiodic phenomena for which there is not evidence of cyclic sensitivity to lighting. *Hourglass* concepts explain this photoperiodism, as in Palmer's description of the photoperiodic control of reproduction in aphids:

> " . . . the aphid *Megoura* . . . produces either virginoparae (parthogenic) or oviparae (sexual) offspring, with the final selection determined by the . . . photoperiod. When kept under very long nights, such as 8 hours of light alternating with 64 hours of darkness, no virginoparae are formed. This trend could be reversed in the usual way by probing for the sensitive spots during the hours of darkness with flashes of light. However, only one responsive site could be located, and it occurred at a time 8 hours after the onset of darkness. If the typical light-sensitivity rhythm had been present, similar peaks at hours 32 and 56 (after dark onset) should have also been located. Therefore, the clock envisioned at work here is something akin to an hourglass that starts running by the transition of light to darkness and runs its course by hour 8, at which point a light-sensitivity moment is produced. The hourglass is not inverted again until the next light-to-dark transition." [250].

Thus, the distinction between the hourglasses and the more complicated explanations involving circadian sensitivity rhythms is made with the resonance experiment.

To discuss these circadian-rhythm-dependent explanations further, we will consider as an example the role of the pineal gland in photoperiodic control of reproduction and some ways that the gland might be mediating photoperiod. A real breakthrough in understanding seasonal reproduction occurred when it was discovered that removing the pineal gland prevented dark-induced regression in hamster testes [178]. This led to further work which substantiated a hypothesis that the pineal gland makes a hormone, melatonin, that inhibits the reproductive system. The importance of this discovery can be seen when we consider the way that melatonin production is regulated by photoperiod. Melatonin production in the pineal gland is controlled by two enzymes (Fig. 6.9). An indole in the pineal gland, serotonin, acts as a substrate for the first enzyme, N-acetyltransferase (NAT), which acetylates it to produce N-acetylserotonin. The N-acetylserotonin in turn is methylated by a second enzyme, hydroxyindole-O-methyltransferase (HIOMT) to form melatonin. The melatonin is released as a hormone and probably achieves its action on the reproductive system indirectly by acting on the hypothalamus of the brain to influence production of prolactin, a gonad-stimulating hormone [44]. The NAT has rhythms that respond to light and dark and thus it controls the daily cycle of melatonin production. HIOMT,

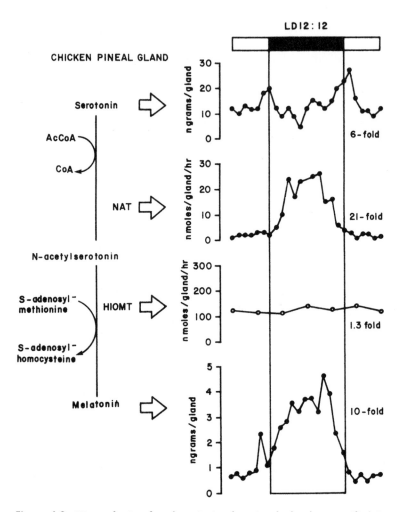

Figure 6.9 Biosynthesis of melatonin in the pineal gland as typified in chicks. Rhythms in the activity of the enzyme, NAT, control cyclic production of melatonin from the substrate, serotonin.

however, also fluctuates in response to lighting and it may modulate the daily cycle of melatonin production.

In fact, the HIOMT was the first pineal enzyme examined for possible control of photoperiodism by its effects on melatonin. The idea was, and data support this, that long nights resulted in greater HIOMT production, in turn larger melatonin production, and thus inhibition of the reproductive system. There are many species (sparrows, chickens, etc.) in which HIOMT does show such a photoperiod-dependent change and could be responsible for seasonal fluctuations in melatonin (Fig. 6.10).

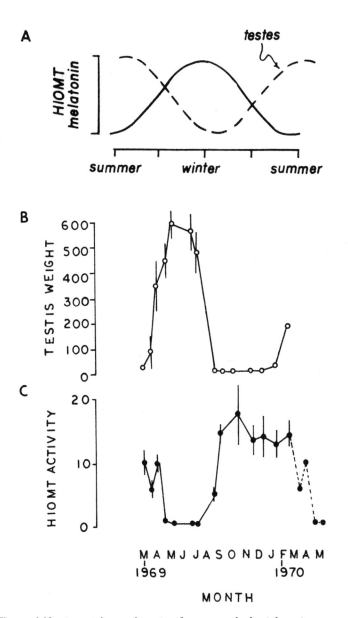

Figure 6.10 An *analog* explanation for seasonal physiology is represented by the response of a pineal enzyme (HIOMT, hydroxyindole-O-methyl-transferase) to photoperiod (A). Here, the idea is that HIOMT amounts are smaller in long days, less gonad-inhibiting melatonin is produced, and testes enlarge (B). The enzyme does show this response in some species such as the house sparrow (C, after Barfuss and Ellis [23]).

Such an explanation was called a seasonal *analog* by Charles Ralph. The HIOMT explanation is an example of a physiological *hourglass*, but does not readily explain the complex responses revealed by resonance experiments which require a circadian sensitivity. However, another enzyme in the pineal does have properties that can explain the light-break and resonance experiments.

When Klein and Weller [198] measured the NAT daily rhythm in the pineals of rats, they found that NAT rose 30-fold at night and that the cycle was suppressed by constant light and persisted in constant dark. Evidence from chickens supported the idea that the melatonin daily rhythm was due to NAT [62]. What is most intriguing is that dark cannot stimulate NAT at just any time, but only during a *sensitive* period that occurs in the night [48] and light causes a rapid decrease (*plummet*) in NAT activity once it is high [199]. With the dramatic responses of NAT to light and dark I have explained [43] light-break and resonance experiments in the photoperiodic control of reproduction (Fig. 6.11). NAT has a freerunning circadian rhythm in constant dark (Fig. 6.11D [46]). The pattern of dark-time NAT depends upon photoperiod (Fig. 6.11B, C; Fig. 6.12). However, in light-break and resonance experiments, a

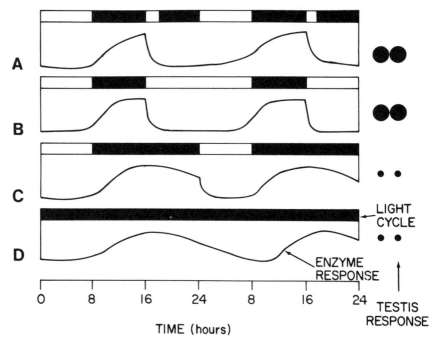

Figure 6.11 A model in which the duration of NAT or melatonin was responsible for measuring daily changes in photoperiod. A particular feature of the model is that the light inhibits NAT when it is high in the dark, and this dark-time sensitivity can account for the light-break and resonance effects because it reduces the length of the perceived dark-time. (Binkley, 1971.)

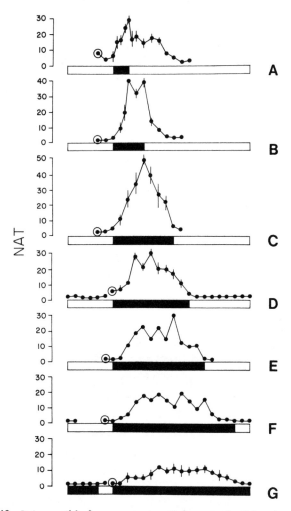

Figure 6.12 It is possible for an organism to keep track of the photoperiod with biochemical events. Here, the response of a chick pineal enzyme (NAT = N-acetyltransferase) is illustrated. The vertical axis represents enzyme activity in nmoles/pineal/h; the horizontal axis is 24h. The bar below each graph illustrates the timing of the light-dark cycle in which the chicks were held to obtain the graph above the bar. The dark-time was shortest (2h) in (A) and longest (22h) in (G). The *time profile* of NAT is a function of photoperiod— the amplitude, duration, and phases of rise and fall of enzyme activity are all a function of photoperiod and of potential use for an organism to synchronize its internal physiology [71].

pulse of light is sufficient to produce a long-day response. NAT accommodates this because a light pulse in the late subjective night suppresses NAT (*rapid plummet*) and NAT cannot reinitiate so that the NAT pattern (and by it the melatonin production) are curtailed as though a short night transpired (Fig. 6.11A) [39]. The data for NAT show that, first, NAT patterns are a function of photoperiod, and second, because of the rapid plummet, NAT can respond to changing daily nightlength. It has been my hypothesis that the D/L taken by the organism as subjective dawn resets the circadian clock to *zero* beginning a new cycle at the commencement of a subjective day.

We note that while the total amount of NAT per night is little affected by photoperiod, the pattern of NAT is altered by photoperiod in a number of ways—the duration of high NAT is longer in long nights, the amplitude of the NAT cycle is higher in short nights, and the phase of NAT increase and decrease alters with respect to any arbitrarily selected phase change [71]. Any or all three of these parameters could be used by organisms as indicators of photoperiod. There is evidence that some animals can use duration of melatonin

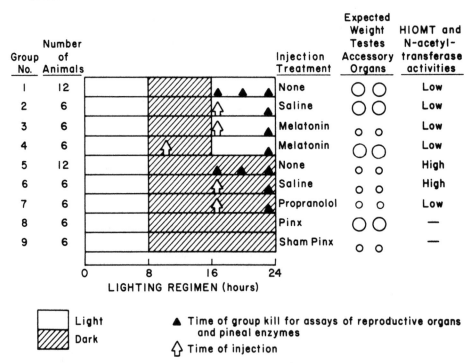

DIAGRAMMATIC REPRESENTATION OF MALE HAMSTER EXPERIMENTS

Group No.	Number of Animals	Lighting Regimen	Injection Treatment	Expected Weight Testes Accessory Organs	HIOMT and N-acetyl-transferase activities
1	12		None	○ ○	Low
2	6		Saline	○ ○	Low
3	6		Melatonin	○ ○	Low
4	6		Melatonin	○ ○	Low
5	12		None	○ ○	High
6	6		Saline	○ ○	High
7	6		Propranolol	○ ○	Low
8	6		Pinx	○ ○	—
9	6		Sham Pinx	○ ○	—

LIGHTING REGIMEN (hours)   0   8   16   24

☐ Light   ▲ Time of group kill for assays of reproductive organs and pineal enzymes
▨ Dark   ⬆ Time of injection

Figure 6.13 Attempts to mimic the photoperiod changes by protocols such as this one for injecting melatonin at various times have supported the idea that duration of melatonin is used to regulate reproductive responses of some species. (Binkley, 1971.)

production as a measure of dark-length for photoperiodic control of reproduction. The first protocols along these lines involved artificial extension of high melatonin by injecting melatonin before dusk or after dawn in a short night (Fig. 6.13).

There has been much discussion of explanations in which photoperiod might be detected by the interaction of two oscillations (in keeping with the hypothetical M and E oscillators mentioned in Chapter 5). The explanations are based on *coincidence*—the phase angle between two oscillators being altered by changing photoperiod. Investigators have defined both *external* and *internal* coincidence.

> "External coincidence. A model for the photoperiodic clock in which light has a dual role: (1) It entrains and hence phase-sets the photoperiodic oscillation, and (2) it controls photoperiodic induction by a temporal coincidence with a photoperiodically-inducible phase." [303]

> "Internal coincidence. A model for the photoperiodic clock in which two or more oscillators are independently phase-set by dawn and dusk, and photoperiodic induction depends on the phase-angle between the two." [303]

The ideas defined in the two types of coincidence incorporate concepts discussed previously. Fig. 6.14 shows how the phase of two oscillations might change with changing photoperiod. It also makes the point that changing the

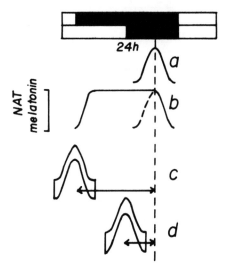

Figure 6.14 Changing the duration of melatonin (A, B) with photoperiod (bars) also automatically changes the phases of its rise and fall with respect to other events—time of day and possible other internal oscillations (open sine wave, C, D). Some investigators think that the phase relation of two oscillators, one of which may be the one that times melatonin production, may be used to trigger reproductive events.

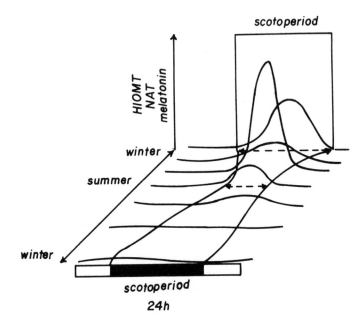

Figure 6.15 In the pineal, two enyzmes (NAT and HIOMT) respond to light, which could produce a response to photoperiod in which duration and amplitude both vary dramatically. The graph represents melatonin in tortoises. (After Vivien-Roels, Arendt, and Bradtke [361].)

duration of one oscillation (e.g., that in NAT and melatonin) changes the phase of the rise and fall of the oscillation (see also Fig. 4.12).

As I pointed out, the duration of NAT is not the only parameter affected by photoperiod—the amplitude of the oscillation also changes. Seasonal patterns of melatonin in the tortoise support the idea of both amplitude and duration modulation of melatonin as the seasons progress. The modulations could be accomplished with responses of NAT alone or in conjunction with changes in HIOMT (Fig. 6.15) [361].

7

CIRCARHYTHMS

"I would like now to discuss some studies which suggest the living clocks to have an extrinsic timing system. I refer to the phenomenon of extrinsic biological rhythmicality observed in so-called constant conditions. This is the occurrence of rhythms whose periods, and at least to some extent, also amplitudes and forms, are derived in direct response to subtle geophysical rhythms. These rhythms, unlike the circadian ones, are *always* present. . . . They even underlie circadian ones when the last are present and appear to merge imperceptibly with the circadian ones whenever the latter synchronize accurately with a natural geophysical period. . . . The circadian rhythms are . . . a consequence of the organisms' normally rhythmic clock-resetting machinery. This means has been termed autophasing."

Frank A. Brown, Jr. [84]

WHAT ARE *CIRCARHYTHMS?*

Aschoff and Pittendrigh [15, 261] categorized rhythms by frequency (period length), of which four *circarhythms* correspond to and synchronize with geophysical cycles in the natural environment (tides, day and night, phases of the moon, and season). The corresponding rhythms are so named: circatidal, circadian, circalunar, and circannual. For these four rhythms, the generalization has been made that they persist, they entrain, and they are temperature compensated [15]. It is the purpose of this chapter to survey these cir-carhythms and other noncircadian rhythms.

ULTRADIAN RHYTHMS

An *ultradian* rhythm is "one with a period somewhat shorter than 24 hours" [250] or with a frequency "greater than 1 cycle in 20h" [168]. The word ultra means beyond and the use of ultra here applies to frequency (the reciprocal of period length). A circadian rhythm has a low frequency ($1/24$ = 0.04 cycles/h), and a cycle with a period near 1h would have a higher frequency ($1/1$ = 1.0 cycles/h). Many examples of ultradian rhythms are available; they are not associated with a geophysical Zeitgeber. Aschoff [15] argued for the reservation of the prefix *circa-* for rhythms that synchronize to a geophysical variable; so if we follow his suggestion, we would not give circa- names to ultradian cycles.

There are many examples of biological processes that have cycles with ultradian frequencies: eyeblinks (24/min in the human) [296], compound action potentials (CAPS, 0–300 impulses/h as in the optic nerve of the *Aplysia* eye) [188], respirations (12–15/min in the resting human), heartbeats (about 70/min in the human), erections (about 90 min cycles in sleeping men), etc. We will consider a few of these here and compare their properties with the properties of circadian rhythms.

First, let us consider the heart rhythm. The pacemaker for the heartbeat resides in a portion of the heart (the sino-atrial node). The pacemaker's signals are transmitted electrically to other parts of the heart. Pieces of hearts are capable of beating when isolated *in vitro*, but the fastest rate is in the sino-atrial node and the signal from the beating node drives the rest of the heart. In the laboratory, beating hearts can be studied in animal preparations. Fig. 7.1 shows recordings from a turtle heart. The beating of the heart can be simply measured by attaching the end of the heart via a thread to a mechanical or electrical device that measures its pull (e.g., a muscle lever and kymograph). The turtle heart beats about 1 cycle/2 sec and the way the recording looks depends on the chart speed of the recorder. Left on its own, the heart expresses its own spontaneous, intrinsic frequency (a *freerun*, Fig. 7.1A), and there is no corresponding geophysical cycle that entrains it. However, the heart can be driven (*entrained*) artificially by electrical signals at a range of frequencies. Physicians make use of implanted electronic pacemakers in order to drive the rate of the human heart. Frequency demultiplication can be demonstrated with the turtle heart by gradually increasing the frequency of stimulation (Fig. 7.1C, D). When this was done, as the rate of signal increased, the heart speeded up and reduced the amplitude of its contraction. When it reached the point where the signal was too fast for it to synchronize, frequency demultiplication occurred and the heart synchronized to only some of the electrical impulses.

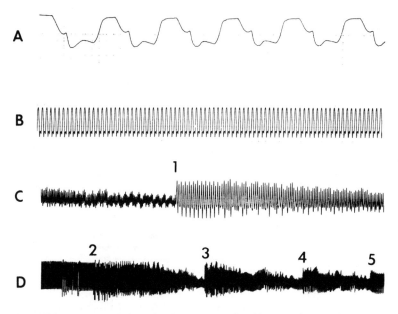

Figure 7.1 Frequency demultiplication can be illustrated with the turtle heartbeat rhythm. Turtle heart contractions (about 30 per minute is the normal rate) are shown with fast chart speed (A) and slower chart speed (B).

The beats can be stimulated electrically. (C) and (D) show slow chart speed records of electrically stimulated turtle heart beats in which the rate of stimulation was gradually and continuously increased. As the stimulation rate increased, the amplitude of contractions decreased, and when the rate became too fast for the heart, the heart slowed down and increased its contraction amplitude. This is an example of frequency demultiplication—five demultiplications (1–5) are shown. (Binkley, 1970.)

Fig. 7.1 shows six frequency demultiplications of the turtle heart contractions as the electrical stimulation rate was increased continuously. Is the heart, then, a biological clock? Hearts do not meet one criterion presumed important by circadian investigators; hearts do not exhibit temperature compensation. That is, the rate of the turtle heartbeat is a function of temperature—it speeds up as temperature increases, it slows down as temperature decreases. Heart-rate is also controlled by neural signals (e.g., signals via the vagus nerve slow the heart down). A transplanted heart is dependent upon its own pacemaker. Superimposed upon the heartbeat rhythm is a circadian rhythm of heart rate (lower at night); characteristics of the donor heart rhythm are claimed to continue in the recipient for transplanted human and hamster hearts or a hamster heart isolated outside the animal [205, 250].

Second, while we are on the subject of hamsters, consider an ultradian rhythm of locomotor activity that we have observed in the Siberian hamster, *Phodopus sungorus* (Fig. 7.2). Each day the hamster sequestered its wheel-running activity into short (usually less than an hour) *bouts*. Most days the

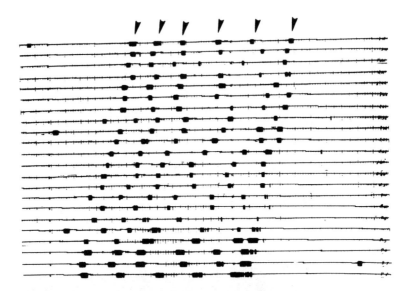

Figure 7.2 Some organisms exhibit a circadian rhythm, but also have short
period components. The record is 22 days of running activity of a Siberian
hamster in DD. Each line is 24h. There is a circadian rhythm with a period less
than 24h. The hamster also divided its daily running activity into short bouts
(e.g., 6 indicated by arrows at the top of the record). (Binkley, Mosher, and
Spangler, previously unpublished.)

hamster exhibited six bouts of activity, but occasionally the hamster exhibited
five or seven bouts. There was also a well-defined circadian rhythm of inactivity. (For another example, see Fig. 2.3) In the case of the hamsters, the
animals were in constant light or dark and there was no geophysical Zeitgeber.

Third, as discussed, sparrows may have a bimodal activity pattern
(Chapters 1 and 3) with peak activity just after dawn and just before dusk. We
have never observed a sparrow in constant conditions that had an ultradian
bout pattern like that above for a hamster. We did observe that it was possible
to drive sparrow-perching activity at ultradian frequencies with short light-
dark cycles (Figs. 3.2, 3.17, 5.5). However, when sparrows were placed in DD
after a short cycle (e.g., LD1.5:1.5), they did not show any evidence of short
cycles, but instead freeran with a circadian rhythm. We pushed this to a
further extreme in a study in which we exposed sparrows to 10 min cycles
(Figs. 7.3 and 7.4) [53]. Sparrows were active in the light of a 10-min cycle with
3 min of light (Fig. 7.3C) producing a *staccato* pattern (Fig. 7.4B). When the 10
min cycle contained only 0.5 min of light, the sparrows were not always active
in the light (Fig. 7.3E); however, the period of the rhythm in the short cycle was
less than the freerunning period in DD (Fig. 7.3F). There was no evidence of
prior 10-min cycles retained in DD. Thus, it is possible for an organism to
follow a cycle for which it does not have an internal frequency; in the case of the
sparrows, this was a direct effect of light. In a similar study using roaches, my
colleague, Shep Roberts, obtained curious results. He found that 30 sec of light

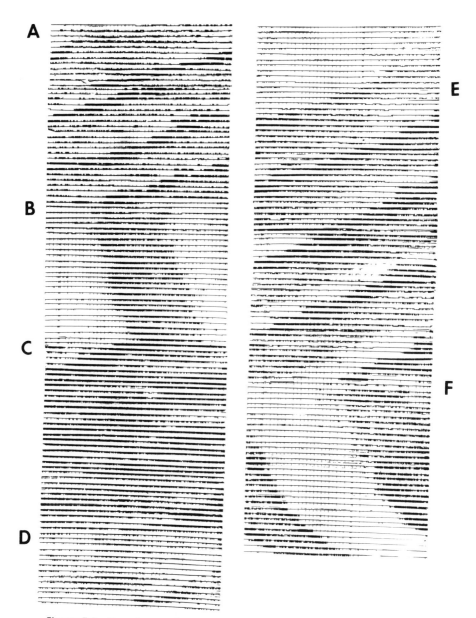

Figure 7.3 Very short cycles, on the order of 10 min, can affect rhythms. The record is from a house sparrow. In (A), the sparrow was in constant light where it exhibited a freerun (period = 22.8h). In (B), the sparrow was placed in DD where it displayed, as expected, a freerun with a longer period length than in LL (period = 24.3h). In (C), the bird was placed in a 10 min cycle in which 3 min of light alternated with 7 min of dark which shortened the freerun; the sparrow was active in 3 min light periods and exhibited a *staccato* pattern (Fig. 7.4). At (D), the sparrow was placed in a 10 min cycle in which half a minute of light alternated with 9.5 min of dark; the light bulb was replaced at (E); the sparrow freeran (period = 22h). In (F), the sparrow was returned to DD where it continued to freerun (period = 24.7h). The short cycles had an effect similar to constant light, they shortened the period length [59].

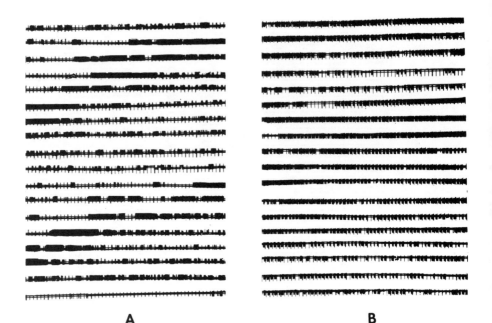

A   **B**

Figure 7.4   Enlargement of the staccato pattern (B) that sparrows showed in
a 10 min cycle with 3 min of light; a same magnification enlargement of LL is
shown for comparison (A). Each line = 10.3h of data [59].

per 10 min cycle shortened the roaches' freerunning period compared to that
in DD. But when the 10-min cycle contained 9.5 min of light, the roaches'
periods lengthened [291].

SLEEP AND 90 MINUTES

Sleep researchers have provided us with an understanding of sleep as an active
process. They have variously divided human sleep into stages [104, 192]. For
example, in one scheme, individuals are considered to be awake, in REM sleep
(rapid eye movement sleep), or in NREM (non-rem sleep includes sleep stages
1–4). Classification of sleep stages is based on recordings from the eye (EOG,
electrooculogram), chin (EMG, electromyogram), and brain (EEG, electroen-
cephalogram). REM sleep is called *paradoxical* sleep because some of its EEG
characteristics are more similar to wakefulness than to other states of sleep. At
the same time it has been more difficult to awaken REM sleepers than individuals in other sleep stages, so that despite the EEG characteristics, REM sleep is
deep sleep. Typically, REM sleep has been characterized by rapid eye movements, dreaming, penile erections, a drop of tonic chin muscle activity, depression of spinal reflexes, and low amplitude EEG waves.

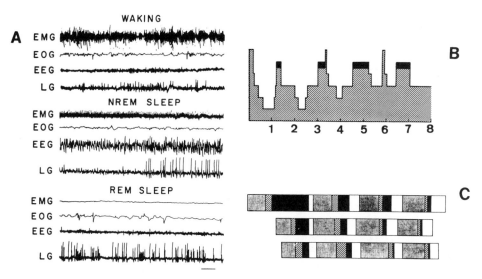

Figure 7.5 The pattern of human REM sleep observed in human brain waves (A) has an ultradian, about 90 min, rhythm represented by the black in the graphs (B,C) showing the distribution of sleep stages for one individual (B) and for three individuals (C). (After figures in Kales [193]). The function of REM sleep is still a subject of debate, but Livermore and Stevens [213] abolished entrainment by light-dark cycles in rats by disconnecting the extraocular muscles responsible for REM.

The occurrence of REM sleep has been mainly documented in mammals, something akin to REM sleep has been observed in birds, and its occurrence in lizards and other lower vertebrates is questionable. The reason we are discussing REM sleep here is that it has an *ultradian rhythm;* in the human it recurs through the night in cycles that are 70–90 min long (Fig. 7.5). In mammals other than humans, the frequency of REM sleep increases as body weight decreases (mice have a 7-min cycle). It has further been proposed that humans have a basic rest-activity cycle (BRAC) for all 24h of which REM sleep is the night-time manifestation. Kleitman described the basic rest-activity cycle:

" . . . BRAC increases in duration, in proportion to body size, in the course of phylogenetic and ontogenetic development. In the rat, the biological hour equals 10 to 13 minutes; in the cat, about 30; in the monkey, 45; in man, about 90; and in the elephant, about 120 minutes. The BRAC lengthens from birth to maturity in all species exhibiting the cycle—in man, from 50–60 minutes in the infant to 85–95 minutes in the adult." [202]

REM sleep is believed to originate from causative factors (neurotransmitters and specific brain regions) in the nervous system, as described by Michel Jouvet:

"The sequence of events leading to REM sleep may be summarized as follows: Priming serotonergic mechanisms that are located in the caudal part of the raphe system act, probably, through deaminated metabolites on cholinergic mechanisms, which in turn (as it was implicated in the peripheral nervous system) may trigger the final noradrenergic mechanism of REM sleep located in the nucleus locus ceruleus." [192]

The processes underlying REM sleep have been investigated with drugs. For example, one drug, para-chlorophenylalanine (PCPA) inhibits synthesis of serotonin in the brain and induces insomnia. It seems probable to me that the cycles of REM sleep and BRACs have a similar innate origin to the cycles of bouts of activity we illustrated for nocturnally active hamsters.

ESTROUS CYCLES

If ultradian rhythms occupy the time domain of rhythms with shorter periods (higher frequencies) than circadian rhythms, then *infradian* rhythms occupy the time domain of rhythms with longer periods (lower frequencies) than circadian rhythms. The rest of this chapter deals with some examples of infradian rhythms. The first of these examples, and the topic of this section, is that of the estrous cycles.

Many animals have reproductive events that recur in an orderly manner, or reproductive rhythms. Probably the ones most studied are the 4–6 day cycles that are found in female laboratory rodents (mice, rats, and hamsters). The animals enter and exit a period of sexual receptiveness (heat, or *estrous*) during which they are willing to mate and able to conceive. Estrous in the rat,

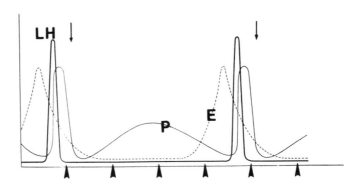

Figure 7.6 Graphs of blood hormone levels (amounts on vertical axis) as they occur in the female rat over the time course of the estrous cycle (horizontal axis). Four to five days (upward-pointing arrows, here four days) elapse between ovulations (downward-pointing arrows). LH = luteinizing hormone from the pituitary, E = estrogen from the ovary; and P = progesterone from the ovary. (After Austin and Short [22].)

for example, occurs during the dark-time for a few hours every fourth day. Correlating with the 4-day heat cycle are other events such as changes in hormone levels and vaginal cytology that recur at 4-day intervals (Fig. 7.6).

Estrous cycles persist in constant conditions. There is not a 4-day Zeitgeber that synchronizes the estrous cycle, but there is a circadian influence so that specific events in the cycle (such as heat) recur at a time of day that can be controlled with a 24h light-dark cycle (Fig. 7.7). Moreover, in those species that have seasonal reproduction, there is also seasonal control over whether the estrous cycle will be present in the female or whether the female reproductive system will be quiescent (just as in the male, Chapter 6).

It is interesting to note that constant light abolishes cyclicity in rodents [1, 91, 370]. In constant light they go into permanent estrous. Removing the ovaries leaves the remaining reproductive system in a quiescent state, diestrous. The hypothalamus of the brain is the likely site of the 4-day clock sending endocrine signals to the ovaries by releasing gonadotrophins (follicle

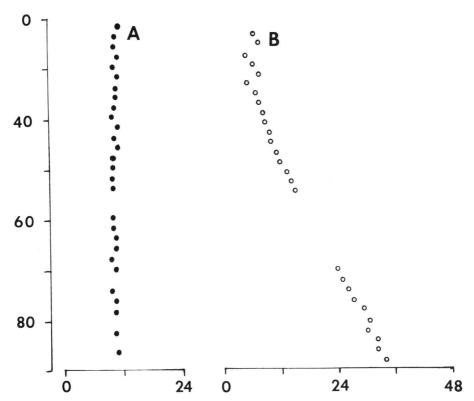

Figure 7.7 Rodents have a 4–6 day reproductive cycle, the estrous cycle, which has a circadian component [1]. Vaginal cells exhibit cornification every fourth day in LD16 : 8 cycles (A); the rhythm freeruns in LL (B). (Redrawn from Alleva [2].)

TABLE 7.1 Estrous Cycles of Some Mammals[1]

Organism	Cycle Length (days)
Cow	21
Goat	20–21
Sheep	16
Pig	21
Horse	19–23
Dog	60
Mink	8–9
Fox	90
Ground squirrel	16
Guinea pig	16
Golden hamster	4
Mouse	4
Rat	4–5

[1]After data in Hadley [166]

stimulating hormone, FSH; luteinizing hormone, LH; and prolactin) from the pituitary gland.

Estrous cycles vary in length depending on the organism (Table 7.1). The female reproductive cycles of diverse organisms also vary in whether the ovulations recur *spontaneously* or whether they are *induced* by stimuli such as mating. Some organisms have continuous estrous (ferret, rabbit) in which ovulation is induced by mating.

LUNAR DAY AND TIDAL RHYTHMS

All the organisms on earth are subjected to cycles from the moon's rotation about the earth. The organisms that live in the seas (Fig. 7.8) have the most easily observed effects because of the tidal cycles that are due to the moon (Fig. 7.9). The lunar *day* is 24.8h (24 hours 51 min) and there are two tides per lunar day recurring at 12.4h intervals. The time between two high tides is called the *tide day*. The longer than 24h period is achieved in the manner described by Palmer:

> "The earth rotates on its axis in a counterclockwise fashion and the moon circles the earth in a similar direction. As the earth completes one rotation (which takes 24 hours) relative to the sun, the moon's travels have placed it in a new location (therefore it is not a stationary reference point). Thus for a given longitude . . . to face the moon again, the earth must catch up by rotating an additional 13 degrees." [250]

Figure 7.8 While circadian rhythms are readily apparent in land-dwelling organisms, organisms that have an ocean environment take on an additional time dimension forced by the tides.

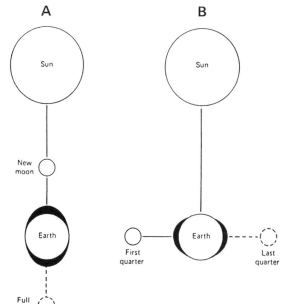

Figure 7.9 The diagram shows how gravitational pulls of the sun and moon produce tides. The filled areas around the earth represent the bulging oceans. (After Palmer [250].)

The amplitude and timing of the rhythm of the tides vary with longitude (large tides towards the poles), phase of the moon, and wind. The rising tide is the *flood* tide; the falling tide is the *ebb* tide; the unusually high tides at the full and new moons are the *spring* tides; the unusually low tides at the first and third quarter moons are the *neap* tides; and a tide running against the wind is a *weather* tide. A beach organism, then, has a number of time cues available that recur in each 24.8h period—dawn, dusk, two high tides, and two low tides.

Under natural conditions, dual rhythms have been measured in some organisms representing the 24h light-dark cycle and the two tides. Figure 7.10 shows the 24h and 24.8h components in the activity of the penultimate-hour crab, *Sesarma* (a nocturnal animal with activity peaking at 11 p.m.). Figure 7.11 shows the freerunning activity of a fiddler crab, *Uca*, in the laboratory in a natural light-dark cycle; note that there are two activity components per lunar day that are believed by some investigators to represent an innate tidal pattern. Another interesting organism, the commuter diatom, *Hantzschia*, has a vertical migration rhythm. The organism is a microscopic alga that inhabits the sand. At high tide, the organism descends among the sand grains, but at low tide, the organism rises to the surface in sufficient numbers to give the beach a golden brown color. Various factors associated with the tides may

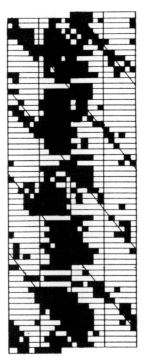

Figure 7.10 Some organisms exhibit both lunar day (24.8h) and solar day (24h) fluctuations. Here, an activity record for the penultimate crab shows the two rhythms. The horizontal axis is 24h; the bars represent days with the filled areas indicating activity; the diagonal lines are drawn through the midpoints of the daily high tides (After Palmer [248, 250].)

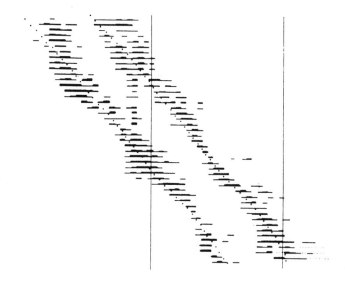

Figure 7.11 The tidal rhythm of a fiddler crab has a daily modulation in natural lighting. Vertical lines are 24h apart; dotted lines represent times of high tide. (After Barnwell [26].)

synchronize the tidal rhythms. Inundation (wetting) does not seem to be important, but there is evidence that temperature, pressure, and mechanical agitation can act as cues in the laboratory.

Obviously, it can be difficult to distinguish experimentally between solar-day (24h) and lunar-day (24.8h) rhythms. For example, in one study [150], a group of sparrows in DD had an average peak freerunning period value of 24.87 +/− 0.54h, which is close to the 24.8h lunar day rhythm and would seem to support a lunar influence. Individual sparrows contributing to the study had a variety of peak freerunning periods (24.78, 24.68, 24.99, 24.86, 24.98, 24.65) which seem to argue against their synchronization by lunar cues in DD in the laboratory. On the other hand, there is presence of bimodal rhythms in entrained sparrows (Chapter 3 and Fig. 1.18) and splitting into two components that occurs in hamsters with a near-tidal day periodicity (Fig. 5.4). If sparrows and hamsters are not somehow picking up lunar signals, then it is tempting to speculate that the occurrence of near lunar-day rhythms in some individuals may reflect ancestral vestiges of timing mechanisms evolved in the sea. In any case, there are many examples of living things (sparrows in constant light with 22h rhythms, for example) with circadian rhythms that are not close to 24.8h.

LUNAR MONTH RHYTHMS AND MENSTRUAL CYCLES

Imposition of a lunar day is not the only periodic signal provided by the moon. There is also a light-dark cycle provided by the waxing and waning of the moon that produces a lunar month (a synodic month, a lunation) of about 29.5 days.

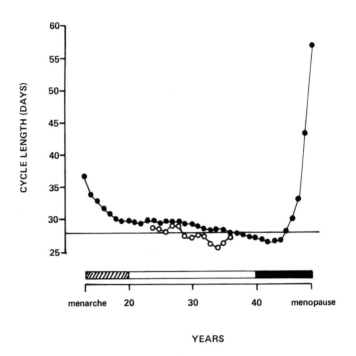

Figure 7.12 The effect of age on the length of the human menstrual cycle from the onset of menstruation (menarche) to its cessation (menopause). (After data of Reilly and Binkley, open circles, and Treolar [342], closed circles).

Monthly variations have been found in biological parameters—e.g., planarian phototaxis and bean water uptake [250]. To the extent that the cycles can synchronize to the lunar month, they comprise *circalunar* rhythms. Probably the most well known monthly cycle is the human menstrual cycle, which does average 29.5 days between the ages of 15 and 40 years (cycles tend to be longer than 28 days at puberty and menopause, Fig. 7.12). We applied the raster method for plotting rhythms to the human female menstrual cycle (Fig. 7.13) and found that the average menstrual cycle length figures cloak the real variability that occurs in menstrual cycle length both within individuals and between individuals (Fig. 7.14).

Is the menstrual cycle a circalunar rhythm? The variation in period length among and within individuals argues against a lunar cue in modern society. However, we mask the lunar cycle with home lighting and window curtains. Dewan [125] attempted to synchronize women's menstrual cycles by keeping on the lights at night for days 14–17 after the first day of the last menstrual cycle and claimed that the treatment regularized the women's cycles toward the 29.5 day average. McClintock argued that reduced individual deviations from the group mean time of menstruation in a college dormi-

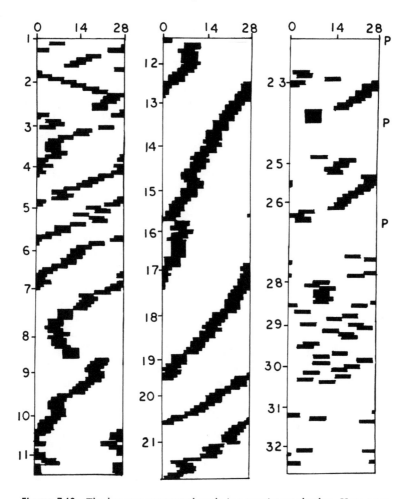

Figure 7.13 The human menstrual cycle is a persistent rhythm. Here a woman's entire menstrual history (from menarche to menopause) has been plotted. The horizontal axis is time in days; the vertical axis is time in years; black bars represent five days of menses; P = pregnancy. When the pattern drifts to the left, the cycle was less than 28 days (most of the record); when it drifts to the right, the cycle was greater than 28 days (e.g., years 2, 8). The rhythm deteriorated with the approach of menopause [276].

tory over the course of a 9-month academic year was evidence for social synchronization of menstrual cycles among women housed together [222]. Even if modern women's menstrual cycles are not synchronized by the moon, one might speculate that the period is derived from an ancestral timing that evolved when our forefathers lived under more natural conditions.

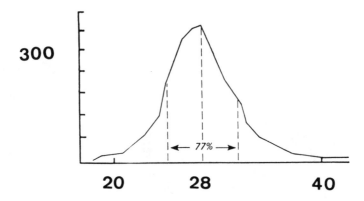

Figure 7.14 The frequency distribution of menstrual cycle length in women shows wide variation. Cycle length (days) is on the horizontal axis and number of cycles of that length is on the vertical axis. The graph represents 2460 cycles from 150 women. (Redrawn [172, 270].)

STILL LONGER CYCLES

Perhaps the most conceptually amazing biological cycles are those that are even longer than a month in length. For example, serotonin fluctuates with two cycles per year in turtles (Fig. 7.15). Such twice-a-year cycles have not been studied much, but a great deal of attention has been paid to annual cycles, the *circannual* rhythms [165]. I discussed seasonal cycles in Chapter 6. In this section I would like to emphasize the fact that some seasonal cycles have been shown to persist in constant conditions. Obviously, demonstration of seasonal rhythm persistence requires considerable persistence by an investigator.

An example of two persistent rhythms are the cycles of fattening and hibernation measured in ground squirrels (Fig. 7.16). When blind ground squirrels, *Citellus lateralis*, were kept at 3°C, Pengelley and Asmundson found that hibernation cycles freeran with periods that ranged from 293 to 369 days in length. Such cycles are evidence of *circannual* rhythms or clocks [165, 253]. Most of the cycles measured were shorter than 365¼ days—the time it takes the earth to complete its circumsolar path about the sun. There are other circannual rhythms besides the feeding and hibernation cycles in squirrels: dry matter production in a plant; growth and longevity in coelenterates; oviposition in mollusks; behavioral thermoregulation in fish; activity in reptiles; molt, migratory restlessness, body weight, fat deposition, and gonad size in birds; and molt, plasma androgens, reproductive condition, nest building, locomotor activity, antler replacement, milk production, and water consumption in mammals [165]. In most of the studies the organisms were held in an unchanging light-dark cycle (e.g., molting was measured in chickadees and warblers kept in LD10:14 for 10 years) [165].

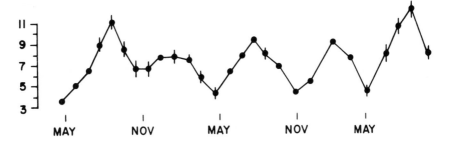

Figure 7.15 There are some rhythms that have two peaks per year. The example shown is for serotonin (ng/pineal on vertical axis) in *Ammocoetes*, the larvae of lampreys. (After Vivien-Roels and Meinel [362].)

Gwinner [165] and Zucker and their colleagues discussed factors that influence circannual period length. First, ambient temperature may affect the duration of the annual cycle of body weight in golden-mantled ground squirrels that had two months longer tau in 9.5°C then they did at 21°C. Gwinner considers studies of beetles, slugs, and seeds to support *temperature independence* for the circannual rhythm. Second, Gwinner considers evidence for a photoperiodic influence upon circannual period length to be equivocal. Third, social isolation reduces the period length of the testicular size cycle in starlings. Fourth, lesions of the SCN or pinealectomy shortened the period in ground squirrels [114, 393, 394]. Fifth, maintaining ground squirrels in 23h or 25h T cycles did not alter the circannual period length from its normal range (e.g., 298–314 days) [92].

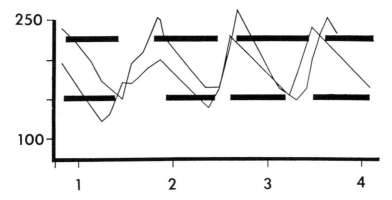

Figure 7.16 Yearly cycles of such physiological events as fattening and hibernation freerun and are called circannual rhythms. Weight (lines, vertical scale in g) and hibernation (horizontal bars) cycles measured in the laboratory are shown for two golden-mantled ground squirrels, *Citellus lateralis*, over a period in excess of three years. (After Pengelley and Asmundson [253, 254].)

Of more interest are studies of potential Zeitgebers for circannual rhythms. First, consider photoperiod. There is all the evidence for photoperiodic regulation of seasonal reproduction (see Chapter 6). Since circannual freeruns are generally not exactly a year, some event(s) must bring the seasonal cycle into synchrony with seasonal events in the environment. Resynchronization was demonstrated (e.g., by transferring woodchucks from Pennsylvania to Australia, thus reversing their annual cues and their body weight cycles). An annual cycle of photoperiod can synchronize a circannual rhythm (e.g., molting in starlings). T cycle experiments have shown that circannual rhythms can synchronize to artificially generated, non-annual cycles (e.g., starlings' circannual rhythms can shorten to as little as 2.4 months and the range of entrainment for the antler replacement cycle of sika deer is about 4–24 months). Second, consider temperature. Temperature changes (e.g., transfer from warm to cold) can phase shift the circannual rhythm of hibernation (e.g., in ground squirrels). Third, consider social cues. Rams may be able to transmit seasonal information to ewes. As you might expect, proving entrainment for a circannual rhythm is an awesome undertaking requiring a special kind of patient investigator, but the work that has been done points to the same kinds of cues in the environment that are used for circadian rhythms.

8

CLOCKSHOPS
AND CLOCKS

"In spite of extensive modeling we have made hardly any progress toward understanding the nature of the physical system responsible for circadian rhythms. All we have are generic models, difficult to disprove which successfully simulate the overt behavior."

Theodosios Pavlidis, 1971 [252]

"All the examples in this book can be presented in terms of one or another hypothetical model of this class (attracting-cycle mechanisms). The trouble with this vision, though, is that it is altogether too easy to construct attracting-cycle models. Such modeling rapidly came into vogue once the idea was broached among circadian physiologists around 1960. The coordinates have different labels in different schemes, and the phase singularity appears in many distinctive guises, but experimental work is not yet refined enough to distinguish between them. At present, the interacting hypothetical quantities, their high-dimensional coordinate space, and its attracting cycle all remain gratuitous constructs, since not one such quantity can presently be named or measured in any particular circadian clock."

Arthur T. Winfree, 1987 [386]

"According to the chronon concept of temporal control proposed here, circadian rhythms in eukaryotic cells and organisms result from the recycling of a sequential machine that regulates the transcription of template RNA from DNA. The eucell is regarded as an event-generator (a clock) whose circadian escapement consists of a sequential transcription component (the chronon) and a chronon recycling (CR) component. The chronon is a very long polycistronic complex of DNA whose transcription rate is limited by some functions of eukaryotic organization that are relatively temperature independent. Each eucell contains hundreds of chronons on each of its nuclear chromosomes, and many sets of extranuclear chronons (free of any attachment to euchromosomes) in its different cell organelles. Upon a given chronon, RNA transcription proceeds unidirectionally from the initiator cistron, C_i, to the terminator cistron, C_t. Once the products of translation of the message of C_t have been formed, the point of no return is passed and the escapement mechanism operates on its CR component. The CR includes post-transcriptional events of translation, end-product formation and polymer assembly, and pretranscriptional events that cause an initiator substance to accumulate; when the initiator arrives at its target cistron C_i, the system proceeds to recycle into its next circadian phase."

C. F. Ehret and E. Trucco [140]

CLOCKSHOPS AND CLOCKS

We can tell the time on a grandfather clock from the position of the *hands* on the clockface. By measuring a circadian rhythm in some variable, we can also tell the time setting; the variable observed is thus analagous to the hands of the grandfather clock. We can open the back and front of the grandfather clock and look directly at the mechanism of wheels and weights that determine the position of the hands; that is, we can look at the clockworks (Fig. 8.1).

Circadian biologists have been intrigued by the *clockworks* that underlie the circadian rhythm. Their efforts to discover the clockworks have taken two

Figure 8.1 We can open the back of a grandfather clock to look at the clockworks. Likewise, we can observe the operation of an hourglass timer. The biological clockworks have been less easy to see.

directions. First, they have treated the clock as a *black box* and have attempted to derive or create stochastic, or conjectural, clock mechanisms from the properties of overt rhythms (measurable oscillating variables) and the rules of physics, chemistry, and mathematics. Palmer gave us a definition of black box as "a container with walls still opaque to the inquiring eyes of science, in which all thus far unexplained processes take place." [250] The black box hypotheses usually take the form of mathematical or physical models. A second approach has been to attempt to locate structures at the organ or cellular levels that either affect circadian rhythms or which have properties that might explain the properties of overt rhythms.

I have collected in this chapter, as mechanical clocks are gathered in a clockshop, a number of the models for the clocks underlying circadian rhythms. The opening quotations of Pavlidis and Winfree, which refer to the black box models, are unsatisfying in that they reflect three decades of heated discussion without much real knowledge of the clock. The information that has been emerging since the late 1960s on the actual locations of circadian pacemakers, their connections with their photoreceptors, is much more satisfying, as is the Ehret model, which is, at least, explicit. Chapter 8 deals with this direction of research, and the specific clocks are discussed with the species in which they reside in Chapter 9.

BLACK BOXES

In this section, I will survey some of the models for biological clocks that have been imagined theoretically by considering the rules of physics, chemistry, and mathematics.

First, there are *hourglass* models. The hourglass is a simple time-measuring device in which sand is funneled by gravity through a small hole from one compartment to another. We still make use of hourglasses as egg timers and to time turns in games. When the volume of sand has completely run through, a measured period of time has elapsed (Fig. 8.1). The term *hourglass* has been applied to similar devices that measure other periods of time and to devices that use other substances than sand (the *waterclock* or *clepsydra* measures time with the flow of water). However, unless someone tips the hourglass to reuse it, the hourglass makes a once only time measurement. Models that are based on the *hourglass principle* have also been proposed for circadian rhythms, and especially for photoperiodism. For example, a photoperiodic hourglass has been proposed for the aphid, *Megoura*, described by Palmer:

> " ... the clock envisioned at work here is something akin to an hourglass that starts running by the transition of light to darkness and runs its course by hour 8, at which point a light-sensitivity moment is produced. The hourglass is not inverted again until the next light-to-dark transition." [250].

Second, another approach has been to compare *pendulum* type oscillators with *relaxation* type oscillators. Wever discusses these oscillators based on a mathematic equation, the Van der Pol equation. According to Wever, the relaxation oscillator "loses the greatest part of its oscillation energy during each period; accordingly, the greatest part of its energy must be replaced from the environment." [377] In the equation, for a relaxation oscillator, the variable in the equation representing energy is much greater than 1. Plots of the solutions of the equation become square-shaped waves as the energy variable increases above 1 (Figs. 8.2, 8.3).

Relaxation oscillators are considered to be nonlinear. A physical example of a relaxation oscillator is the siphon in a tank filling with water—when the water level reaches the top of the siphon (representing the environmental energy input), the syphon empties the tank quickly, depending on its diameter. As pointed out by Mercer [233], relaxation oscillators are readily triggered (entrained) by adding water or by applying suction to the siphon. A pipet washer is made to oscillate this way: water trickles continuously from a tap

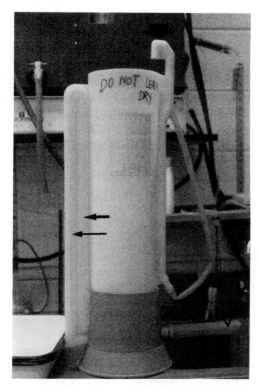

Figure 8.2 A pipet washer is a physical example of a relaxation oscillator. Water is tubed from a faucet so that it runs continuously into the top of the washer (heavy arrow). As the washer fills, so does a pipe (medium arrow). When the water level reaches the top of the pipe, it overflows into another pipe (fine arrow) so that a siphon is created. The siphon, which is larger diameter than the input tube, siphons out the water from the washer at a faster rate than it is coming in from the faucet; the water exits through a large diameter tube (open arrow). Once the washer is emptied, the siphon is interrupted and the washer is again filled by the water from the faucet. As long as water continues to come from the faucet, the washer oscillates between filling and emptying. The rate of oscillation can be controlled by altering the flow of water from the faucet.

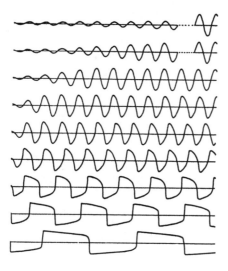

Figure 8.3 The graphs show the solutions of the Van der Pol equation producing a continuum from a pendulum to a relaxation oscillation. The lines represent solutions of the equation with differing amounts of energy exchange between the environment and the oscillation. At the top, the lines represent a pendulum type, sinusoidal, oscillation with a small energy exchange; more time (horizontal axis) is required to reach a steady state amplitude. At the bottom, the graphs represent a relaxation type, rectangular, oscillation with a large energy exchange; steady state amplitude is reached immediately. (Redrawn after Wever [377].)

into a tank; each time the water level in the tank reaches the top of a siphon in the tank, the water runs out. The system oscillates because the water runs out of the siphon at a faster rate than it drips in from the tap. The other extreme case based on the Van der Pol equation is that of the *pendulum oscillator*. Here the amount of energy exchanged between the oscillator and the environment is small, the energy variable in the equation is less than 1. An example of a pendulum type of oscillator is a child's swing (Fig. 8.4). Only a small amount of energy is required by the child to keep this recreational pendulum swinging. When the solution of the Van der Pol equation is plotted for various energy inputs, the oscillations are sinusoidal shaped. The period of the pendulum is very stable. Characteristics of relaxation and pendulum oscillators are compared in Table 8.1 using terminology related to circadian rhythms.

To illustrate the use of the models, consider the characteristics of sparrow perching rhythms with respect to aspects of pendulum and relaxation oscillators in Table 8.1. First, sparrows entrain to light cycles; in fact, as other organisms, to very dim light-dark cycles. It would seem that little energy input from the environment is needed. The range of entrainment, in excess of 16–29 hours (Chapter 3), ($5/24$ or $8/24$, less than half of a cycle) seems small. The sparrows do show frequency demultiplication in 2-pulse skeletons but not in LD3:3, for example. The freerunning period is only slightly dependent on the

Figure 8.4 Some models for biological clocks have been based on the properties of a pendulum. The swing is a pendulum which, by itself, will swing (oscillate) with a constant period until its motion dies down. By pumping, the child can maintain (entrain) the motion of the swing.

TABLE 8.1   Comparison of Pendulum and Relaxation Oscillators[1]

Variable	Pendulum	Relaxation
Energy input from environment to entrain	small	large
Range of entrainment	small	large
Frequency demultiplication	less	more
Dependence of freerun period (or frequency) on environment	less	more
Range[2] of oscillation	varies greatly	varies little
Effect of varying oscillation mean on frequency	no effect	changes greatly
Effect of varying oscillation mean on range	varies	constant
Shape of oscillation[3]	more sinusoidal	more square
Linearity	linear	nonlinear
Effect of increasing mean of oscillation on duration of positive part of oscillation[4]	no effect	increased

[1]derived from texts of Wever and Mercer [233, 377]
[2]range = maximum − minimum value
[3]shape of oscillation, Chapter 1
[4]this relates to level-threshold hypothesis discussed in Chapter 8

environment (LL shortens the freerun from 25h to 22h). We have no information on the oscillation mean or amplitude for perching. The shape of the
freerunning oscillation is more sinusoidal. If my assessments from perching
records (which are the *hands* of the clock, to some extent removed from the
clock itself) are meaningful for distinguishing oscillation types, sparrow circadian rhythms have more pendulum oscillator properties. Appropriately,
Wever cautions against making conclusions from the *shapes* of oscillations:

> "In general, the shape of an oscillation is not suitable criterion for identifying
> pendulum oscillations and relaxation oscillations. Relaxation oscillations can
> assume many shapes other than a rectangular one, for example, a saw tooth
> shape, or even a sinusoidal shape, which might be mistaken as a pendulum
> oscillation. Furthermore, it is impossible to determine from the observed shape of
> a rhythmic function the true shape of the underlying oscillation. In a biological
> system, an unknown number of other functions are interposed between a funda-
> mental oscillation and the observed function." [377]

With these reservations in mind, I think it is still interesting to look at the
effects of lighting on the shapes of the rhythms measured for sparrows (perching rhythms and PRCs). I had been trying to explain the bracket-shaped
autocorrelation graphs I obtained in sparrows. Most sparrows have night

activity so that their freerunning rhythms were skewed sinusoidal waves (Fig. 1.19), though a more square-shaped freerun was measured in a bird with clear onsets. However, when the sparrows are placed in LD, their perching patterns become more square-shaped or, in many, the patterns are bimodal (Figs. 1.18, 3.11). It was quite simple to mathematically generate sine waves and square waves with a computer. I was also able to add them together in various combinations. When I added a truncated sine wave (the below mean half of one sine wave) to a square wave, I obtained a bimodal pattern (remarkably like a sparrow's) that had a bracket-shaped autocorrelation function (like a sparrow's) as shown in Fig. 8.5. From this, I suggest that a description of the sparrow's perching circadian system is as a pendulum-like oscillation upon which light-dark cycles superimpose a square wave, producing more relaxation oscillator-like properties. I attribute the bimodal pattern to the relative actions of the direct effect of light (light causes hopping) and the sparrow's innate circadian rhythm. Even so, the reader will see that after a long discussion, I am little closer to describing the *clockworks* that comprise the circadian pacemaker of sparrows.

Third, some of the circadian theoreticians have discussed a model of the clock visualized as a *limit cycle* or *attracting limit cycle* [149, 252, 383, 384, 386]. In their models, the circadian clock is viewed as a loop (obtained when pendulum velocity is plotted against position, or any two variables, x and y, Fig. 8.6). A cycle of a freerun represents a course, or *trajectory*, around this loop.

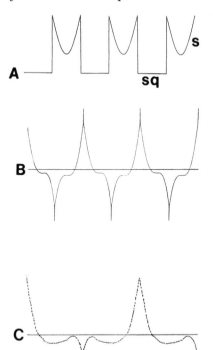

Figure 8.5 The figure shows how a bimodal pattern can be explained mathematically and the consequences for autocorrelation analysis. In (A), a square wave (sq), representing the light-dark cycle, was added to half of a sine wave (s), representing the freerun, to produce a bimodal pattern, much like that observed for sparrows (see Fig. 1.18). When the waveform in (A) is subjected to autocorrelation, the bracket shaped graph shown in (B) results. An autocorrelation for actual data from a sparrow kept in LD12 : 12 is shown in (C).

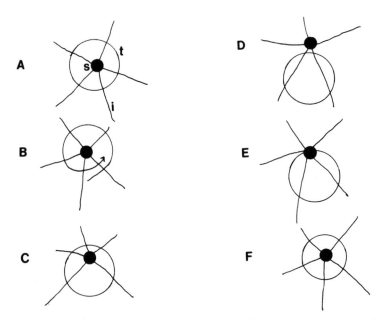

Figure 8.6 Representations of the circadian clock as an *attracting limit cycle.*
The circling line (t) represents the *trajectory;* time positions on the loop are the
isochrons (i); the central dot is the *singularity* (s). (A) represents freerun. (B) is
starting from LL/DD. In (C), a weak stimulus changes the position of the clock
with respect to the singularity (a small phase shift, or *odd resetting*). In (D), a
strong stimulus moves the trajectory so far that it no longer encloses the
singularity, and only can traverse part of the isochrons (a large phase shift, or
even resetting). (E) depicts the case where a stimulus (T*S*) moves the trajectory so that it intersects the singularity—in this case it is possible to proceed
directly to any and every isochron (arrhythmia or unpredictable phase shift).
(F) is the same as (A) but with a smaller loop diameter representing a change in
velocity (as occurs when light intensity changes). (After Winfree [386].)

Time positions on the loop (lines from a *center* or *singularity* crossing positions
on the circle) are called *isochrons* (Fig. 8.6A). An LL/DD transition can start a
clock stopped (e.g., at c.t. 12) on its trajectory (Fig. 8.6B). Perturbing stimuli
(e.g., light pulses, Fig. 8.6C, D) move the position of the clock along its trajectory with respect to the singularity, producing resetting. If the clock is forced
to the center of the circle, the result is arrhythmia or a jump to an unpredictable new phase (Fig. 8.6E). Outside forces (e.g., changing the intensity of
constant lighting, Fig. 8.6F) can change the diameter of the circle. If the system
is more complex than described by a perfect circle, the shape of the loop may be
distorted so that it is not a circle, the convergence point for the isochrons may
not be the center, and the length of the trajectory can be changed by outside
forces (e.g., light intensity of dim LL).

Fourth, a number of investigators have made computer simulations [148,
236, 315]. These simulations are based on characteristics of circadian

rhythms, and these simulations have some success in predicting circadian responses. For example, Enright [148] has made simulations based on nocturnal hamsters and compared those with simulations based on diurnal birds. As Wever summarizes:

> "A sound agreement between model predictions and experimentally derived properties of circadian rhythms indicates that simple laws of oscillation theory govern the apparently very complex rhythmicity in the behavior of living organisms, including humans. It is tempting to conclude from this result that the basic structural mechanisms of generating and controlling circadian rhythms are also simple; and the evaluated dynamics of biological systems should assist in discovering these mechanisms." [380]

COUPLING AND CIRCADIAN ORGANIZATION

An approach to explaining how circadian rhythms are timed has taken more biological directions than the mathematical-physical type models (Chapter 8). To some extent, the ideas that are surveyed in this section are still *black box* models but they are based more on biological processes.

First, *single cells* are probably capable of generating circadian rhythms. This follows from several lines of evidence. (1) Circadian rhythms have been studied (swimming activity, phototactic sensitivity, bioluminescence, photosynthetic capacity, sexual reactivity) in populations of unicellular organisms such as *Euglena, Gonyaulax, Paramecium,* and *Chlamydomonas* [90]. However unlikely, it is possible that these populations of individual separate cells somehow produce circadian rhythms by a mutual interaction. (2) Therefore, another approach has been a sometimes heroic effort to prove that isolated single cells can generate circadian rhythms. The evidence from these attempts is the main underpinning for the belief of circadian investigators that some single cells contain the entire mechanism of the circadian biological clock. Sweeney [319] presented evidence for rhythms of photosynthetic capacity by measuring oxygen production of single cells of *Gonyaulax polyedra* in LD and LL. Vanden Driessche and others have studied rhythms in the single large cells of *Acetabularia* [355]. Deguchi and Takahashi [123, 293] have measured rhythms from chick pineal cells isolated in cell cultures. Taken together, the results mean that there must exist a mechanism for circadian pacemaking requiring only the components available within a single cell.

Second, many organisms are not unicellular but are composed of many cells that may be organized into various tissues; the various components of multicellular organisms may be *coupled.* Internal synchrony could be achieved if each cell of a multicellular organism possessed a circadian clock and the necessary sensors to detect Zeitgebers. However, while independent synchronization may occur in several organs, there is evidence for interaction among internal oscillators. If a process can be controlled by a pacemaker within an

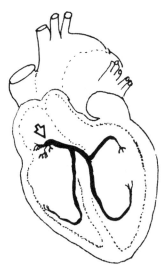

Figure 8.7 The pacemaking system of the human heart. The heartbeat is generated by the cardiac conduction system consisting of the sinoatrial node (arrow), internodal atrial pathways, atrioventricular node, Bundle of His, and the Purkinje system. All of the parts are capable of spontaneous discharges that can make the heart beat, but normally the sinoatrial node acts as a pacemaker whose signal synchronizes the activity of the rest of the system. (After Ganong [161].)

organism, it is referred to as *coupled* to the clock [90]. When two rhythms have very precise synchronization, the *coupling* is *tight;* when the timing of two rhythms with respect to each other is imprecise, the coupling is *loose.* If a treatment causes two rhythms to freerun independently of each other, they have become *uncoupled.* When one oscillation controls another, the controlling oscillator is designated the *pacemaker* or *driving oscillation* and the regulated oscillation is called the *slave* or *driven* oscillation. In making the terminology, circadian biologists borrowed concepts from cardiac physiology (Fig. 8.7).

Third, the coupling mechanisms for circadian rhythm internal synchronization currently emerging have been organized into *hierarchies* of oscillators and driven oscillations (Fig. 8.8) [238]. For such a hierarchy, the environmental time cues (Zeitgebers, e.g., light) are detected (e.g., by the photoreceptors). The environmental information is then relayed by some means (e.g., nerves) to a pacemaker (e.g., a group of cells) which in turn drives measurable oscillations and/or passive responses (cells or processes incapable of self-generated rhythms but able to respond to rhythmic signals). In the hierarchies, the coupling mechanisms might be any of a variety of possible physiological means—nerve tracts, hormones, or temperature.

Fourth, a popular view of circadian organization involves various *two oscillator models* [238, 260]. For instance, in typical two-oscillator models, there is one oscillator that is *locked* to (synchronized by) dawn, an M (morning) oscillator. A second oscillator is locked to dusk, the E (evening) oscillator. These models seem especially appropriate for circadian explanations of photoperiodism, bimodal rhythms, and splitting (Chapter 6). The models account for the changes in relative timing of two parameters in response to photoperiod (e.g., the rise of NAT at dusk, and the fall of NAT before dawn, see Fig. 6.14).

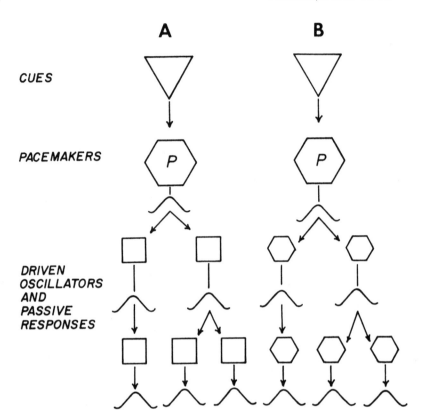

Figure 8.8 Most investigators believe that circadian physiology is a result of a hierarchical organization. Two are shown here (A, B). Each species may have its own hierarchy specialized for adapting to the time constraints of its particular environment.

    At the top of the hierarchy are the organs that detect the Zeitgebers (e.g., the eyes or extraretinal light receptors).

    The pacemakers, (P), form the next level of the hierarchy. These organs (such as the pineal and SCN) are self-sustained oscillators—that is, in and of themselves they can generate a circadian signal.

    The pacemakers may drive other self-sustained oscillators (indicated by hexagons), or may signal processes which themselves are not capable of producing an oscillation (squares).

    The connections between the components of a given hierarchy (vertical lines and arrows) are nerve pathways (e.g., the retinohypothalamic tract) or hormones (e.g., melatonin from the pineal) [238].

    Fifth, it is also possible to account for changes in the shape of an oscillation with a *one oscillator model* such as that proposed by Wever (Fig. 8.9) [376]. In the model, a circadian oscillation has a *mean* (average) value that interacts with a threshold in such a way that an observed rhythm is determined by how much of the oscillation is above the threshold value. By raising and lowering

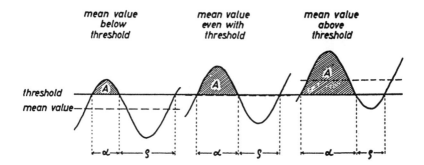

Figure 8.9 One oscillator model in which the oscillation is represented by a sine wave. The observed oscillation is represented by the hatching. The sine wave has a mean value (dashed line). If the *mean value* of the oscillation is changed with respect to a *threshold*, the observed oscillation changes. For example, the mean is lower than the threshold on the left so that the observed oscillation is small (short alpha, as occurs in birds in DD); but on the right, the mean of the oscillation is above the threshold (longer alpha, as occurs in birds in LL). (After Wever [376].)

either the threshold or the mean, the amount above threshold is altered and with it the duration and amplitude of the portion above the threshold. The model is particularly useful for explaining the effects of light intensity on alpha/rho ratio (Chapter 2) and the change in duration of an oscillation (e.g., of day-time activity, night-time NAT, or PRC shape) in response to photoperiod.

ANATOMICAL AND BIOCHEMICAL CRITERIA

The *models* discussed to this point for biological clocks are all hypothetical schemata, imaginings or stochastic schemes, however clever, based on the observed properties of circadian rhythms. Attempts have been made to find visible or chemical clockparts. The two approaches to locating the clockworks have been anatomical and biochemical. In this section, I will discuss these approaches and give some examples. In the next section and in Chapter 9, the reader will find more on the subject of actual biological mechanisms.

The *ablation* (e.g., removing a suspected area with surgery or lesions) approach has been successful in identifying potential pacemakers in multi-cellular organisms. The route of investigation that has been particularly fruitful has been to identify the Zeitgeber detectors (e.g., photoreceptors) and to trace their connections inward to find the pacemakers. In some cases, the photoreceptors themselves are capable of generating circadian oscillations. As an example, the ablation method is outlined in Fig. 8.10 for regulation of the pineal NAT rhythm of the rat.

The general idea is to pick likely structures for photoreceptors or pacemakers, then to ablate them, usually surgically, and then to measure a circa-

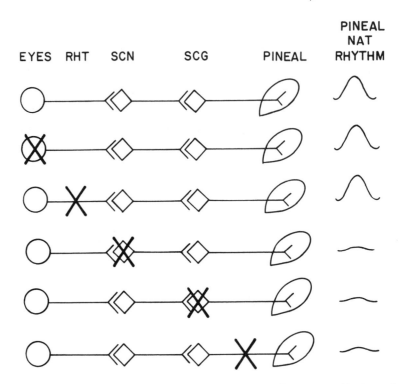

Figure 8.10 Ablation experiments have been used to find potential biological clocks and to discover their connections with photoreceptors. The diagram illustrates ablations used to test connections for the suprachiasmatic nuclei. An organ (e.g., the SCN) was removed or lesioned, or a neural connection was cut (e.g., the retinohypothalamic tract, RHT) followed by measurement of a rhythm (e.g., pineal NAT). When a rhythm (represented by a sine wave) was extinguished (as in the lower three procedures), it implied that the discon-nected structure was either a biological clock, or that the structures formed a connection between that clock and the measured rhythmic variable.

dian rhythm with a Zeitgeber cycle and in constant conditions. If the ability to entrain (e.g., to an LD cycle) is eliminated, but the freerunning rhythm is detected, the structure is a candidate for a Zeitgeber detector (e.g., photorecep-tor). If arrhythmia results, the structure is a candidate for a slave oscillator, or a coupling device, or, most interesting, a pacemaker. In the example of the rat (the evidence for which I have reviewed [40]), blinding abolishes entrainment to LD14:10 [198] so the eyes are the photoreceptors. Lesioning the supra-chiasmatic nuclei (SCN) in the hypothalamus eliminates the rhythm, so it is the candidate for the circadian pacemaker. Cutting various nerve tracts (e.g., RHT) or removing the superior cervical ganglia (SCG) eliminates the pineal NAT rhythm so the neural pathways are viewed as neural couplers. We can make a hierarchy (as in Fig. 8.10) for the rats. The eyes are the photoreceptors, the SCN is the circadian pacemaker, the pineal NAT enzyme is a driven circadian rhythm, and the pineal hormone melatonin (produced in a circadian rhythm by the pineal due to the action of NAT) can carry rhythmic signals to drive further slave circadian oscillations and passive responses. In order to further qualify as a Zeitgeber detector, the candidate structure must have anatomical or chemical means of detecting the kind of signal involved (e.g., the photoreceptor should possess structures or chemistry characteristic of photoreceptors such as lamellar membrane organelles and photopigments such as rhodopsin). In order to further qualify as a pacemaker (Table 8.2), an isolated oscillator should be able to generate a circadian rhythm and to restore rhythm and phase when transplanted. Pacemaking rhythms should be able to synchronize slaves. It is also logical that pacemaker rhythms should reset more rapidly (instantaneous resetting) with fewer or no transients as compared to the slave oscillations they drive.

The chemical or drug approach has been used to study circadian rhythms in organisms, tissues, or cells. In doing this, there is always the logical uncertainty whether any change obtained in an experiment is due to alteration of the pacemaker itself, to changes in the coupling mechanism, or to modification of an observed rhythm. A thought experiment illustrates this: If the feet of a sparrow were amputated, its perching rhythm would be abolished, but it would be an error to conclude that the sparrow's clock was in its feet. Nonetheless, there have been many chemical experiments and information about the circadian clock can be gleaned from them. Table 8.3 classifies the types of responses that have been found.

Several investigators have attempted to entrain circadian rhythms with repeated injections. The problem with the experiments is that often the organism entrains to control treatments (e.g., saline injections), presumably be-

---

TABLE 8.2

CRITERIA FOR PHOTORECEPTOR
ablation blocks entrainment to LD cycles
presence of photoreceptor-like structures
presence of photopigment

CRITERIA FOR PACEMAKER

drives phase/period of other oscillations	correlates with and explains:
rhythm persists in isolated structure	responses to skeletons
transplant carries rhythm/phase	LL attenuation
ablation produces arrhythmia	alpha/rho ratio
link to photoreceptor	entrainment
rapid resetting	photoperiod
	resetting
	freerun

TABLE 8.3

CRITERIA FOR DRUG EFFECTS
synchronization to repeated treatments
freerunning period alteration
amplitude attenuation
arrhythmia
resetting

cause of the disturbance of the injection procedure. However, some success has been obtained with this technique using the hormone melatonin (Fig. 8.11). Rats entrain to repeated melatonin injections. Once synchronized, they commence activity at the time of the injection; this is satisfying because that is the time that melatonin would normally be increasing in the nocturnal animals [101, 275]. The result is made more interesting because daily melatonin injections also synchronize other species such as lizards, and the diurnal lizards begin their rest period at the time of the injection, which again correlates with the timing of the lizards' normal nightly onset of melatonin production [352]. Single melatonin injections cause sparrows to assume a roosting posture (typical of night-time rest) and to lower their body temperature (which also normally occurs at night) [32]. New technology, permitting cyclic drug

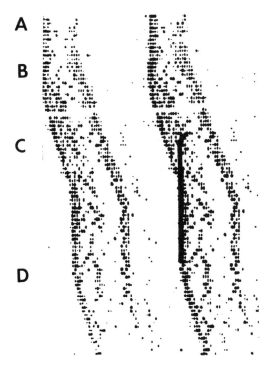

Figure 8.11 Repeated melatonin injections can synchronize the circadian rhythm of an organism. Here, a rat entrained to a light-dark cycle (A) or freeran in constant dark (B), (D). At (C), the rat received daily melatonin injections in constant dark at the time marked by the vertical line—the rat synchronized its activity onset with the injection. Melatonin would normally increase at the time of the rat's activity onset at lights-out. (After Chesworth, Cassone, and Armstrong, [101].)

administration without disturbing the organism, should allow an increase in studies of synchronization to cyclic drug administration (Chapter 10). The results of such studies do not prove that the chemical (e.g., melatonin) is a natural synchronizer, but show only that it is possible for the organism to use that signal.

Alteration of freerunning period has been considered to be evidence that a chemical actually interferes with or alters the circadian pacemaker since the period length is believed to be derived from the pacemaker. There are some chemicals that do alter period lengths (Table 8.4). An interesting aspect of these studies is that the period lengthening persisted for some time after the chemical stimuli were removed. Theophylline and theobromine are methyl xanthines, chemical relatives of a commonly used substance, caffeine. Considering the length of time that scientists have been investigating circadian rhythms and the considerable interest in drugs and pharmacology, the list of period-altering drugs seems sparse. Pittendrigh and Caldarola have argued for a sort of chemical compensation in the matter of circadian period length ("homeostasis of the frequency of circadian oscillators"); that is, for a biological clock to work, its period must be protected from assaults by changes in the environment (temperature, chemical, etc.).

" . . . if circadian oscillations are to function as a reliable framework for a temporal organization of cellular function relative to sidereal time, their . . . period . . . must be essentially invariant in the face of all variations they encounter in the cellular milieu . . . Our proposition is that the frequency of circadian oscillators

TABLE 8.4   Chemical Alteration of Circadian Period Length[1]

Chemical	Species	Rhythm	Effect on period[2]
estradiol	*Mesocricetus*	activity	−7 min
0.1% ethyl alcohol	*Gonyaulax*	bioluminescence	−
0.2% ethyl alcohol	*Phaseolus*	leaf movement	+2−4h
0.5% ethyl alcohol	*Excirolana*	activity	+1h
0.1% theobromine	*Phaseolus*	leaf movement	+2−4h
0.1% theophylline	*Phaseolus*	leaf movement	+2−4h
heavy water, D_2O	*Euglena*	activity	+4h
heavy water, D_2O	*Phaseolus*	leaf movement	+6h
heavy water, D_2O	*Leucophaea*	activity	+1h
heavy water, D_2O	mammals	activity	+1h
lithium	*Kalanchoe*		+
colchicine			+/−
urethane			+/−

[1][90, 146, 240, 250, 338]
[2]+ lengthened, − shortened

is subject to general homeostatic control in the face of all potential perturbations normally encountered in the cell." [263]

Reduction of amplitude by a chemical treatment probably only means that the measured process has been suppressed [388]. The extreme of such suppression is the production of arrhythmia, which may or may not represent *stopping* the clock. Some substances that have suppressed rhythms include: dichlorophenyl dimethyl urea, puromycin, NaCN, 2,4-dinitrophenol, NaF, arsenate, cyanide, and actinomycin D [90]. Bünning argues that suppression of the rhythm (without phase shifting and/or period change) is not evidence that the pacemaker itself is the subject of the tampering. In Chapter 5 (Fig. 5.10) are data showing arrhythmia produced with constant melatonin administration via implanted capsules or drinking water [54, 347]; the effect was immediate, reversible, and dose dependent.

There is a greater profusion of examples (Table 8.5) of chemicals causing advances or delays (resets) of circadian rhythms in response to a single chemical treatment (pulse). There are even some phase response curves that have been measured for chemical perturbations. Some investigators have taken alteration of the PRC as evidence that a treatment has tampered with the pacemaker, which stems from an implied belief that the PRC represents the status of the pacemaker.

CELLULAR CLOCKWORKS

Many investigators have put forward models for circadian pacemaking that are based on cells, cellular components, or molecular biochemical events within cells.

First, because of correlations between cell division and circadian rhythms, a *cell cycle clock* or *cytochron* was proposed by Edmunds [138, 250, 325]. Circadian rhythms in cell division have been a subject of investigation, for example, in *Euglena* (Chapter 9). Cell biologists have identified the events of cell division, or mitosis, and have described a cell cycle with a sequence of four classical events:

$$G_1 = \text{gap 1}$$
$$S = \text{synthetic, chromatin replication}$$
$$G_2 = \text{gap 2}$$
$$M = \text{mitotic, mitosis}$$

Cells (*Chlamydomonas, Chlorella, Euglena, Gonyaulax, Pyrocystis*, etc.) have persistent rhythms of cell division with division (M) occurring in the subjective night. It follows from the correlation of the cell cycles with circadian rhythms that either the cell cycles derive their timing from the circadian pacemaker or that the cell cycle is the basis of the circadian clock (Fig. 8.12). In

TABLE 8.5 Phase-Shifting Circadian Rhythms with Chemicals[1]

Chemical	Organism	Rhythm
alcohol [338]	*Gonyaulax*	glow
anisomycin [321]	*Mesocricetus*	activity
anisomycin [337]	*Gonyaulax*	glow
cGMP [153]	*Aplysia*	optic nerve
cycloheximide [126]	*Gonyaulax*	glow
forskolin [151]	*Aplysia*	optic nerve
glutamate [227]	*Mesocricetus*	SCN
melatonin [350]	*Sceloporus*	activity
serotonin [151,152]	*Aplysia*	optic nerve
serotonin [247]	*Leucophaea*	activity
valinomycin [218]	*Gonyaulax*	glow

[1]Chemicals listed are some of those for which a PRC was obtained. Some authors comment on the drugs used that had no effect [90, 175, 263].

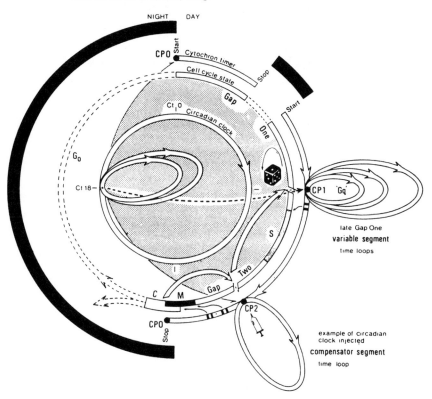

Figure 8.12 Representation of the cell cycle clock for the cell division cycle of *Euglena*. (From Edmunds, with permission [138].)

the model, a *cytochron timer* starts at lights-on and the classical stages G_1, S, and G_2 proceed and trigger M and cytokinesis. Other aspects have been inserted into the model that *program* the clock in the light-time and account for "entrainability, persistence, initiation, phase shiftability, and temperature compensation ... and a singularity point" [138].

Second, Ehret and Trucco proposed a *chronon model* for the circadian clock that makes use of the previously discovered events by which proteins are synthesized in cells (Fig. 8.13) [140, 250]. The chronon model is based on the process of protein synthesis in cells. A sequence of chemical *bases* in the genetic material in chromosomes of the nucleus (DNA) specifies a sequence of complementary bases in the formation of a large molecule (RNA), a process called *transcription*. In turn, the RNA leaves the cell nucleus and provides instructions for forming the sequence of amino acids in proteins at the ribosomes, a process called *translation*. Ehret's idea is that a cell contains *chronons*, strands of DNA 200–2000 cistrons long, which are the rate-limiting components for transcription [250].

$$\frac{\text{Chronon DNA template distance (nm)}}{\text{RNA transcription rate (nm/h)}} = 24\text{h}$$

Thus, the transcription process takes a certain duration of time. At the end of the process, some substance synthesized at the ribosomes then diffuses back to the DNA to restart the cycle of transcription. The role of diffusion in this model provides an explanation for temperature compensation. The model has predictions listed by Palmer.

1. the ... longest chronons ... should be of ... equal length ...
2. circadian clocks should be confined to eukaryotic cells ...
3. ... impediments to diffusion should slow down the sequential transcription component ...

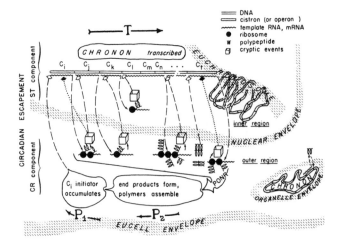

Figure 8.13 Representation of the chronon model. (From Ehret and Trucco, with permission [140].)

4. different mRNAs should be synthesized at different times of the circadian cycle . . .

5. enzymatic activities and metabolic related functions should also display a temporal phenotype . . .

6. correlations (should be found) between the events and mechanisms that regulate the circadian cycle . . . and the cell (mitotic) cycle [250]

As pointed out by Palmer, many of these predictions are met (especially 2, 3, 5, 6).

Third, the *membrane model* of Hastings and co-workers for the circadian clock is based on the structures of membranes, the characteristics of potassium ions, and the circadian Q_{10} evidence for a diffusion process (Fig. 8.14) [242, 250, 322]. Cell membranes are viewed as consisting of a lipid bilayer studded with moveable proteins. Changes in this "fluid mosaic membrane" [242, 243, 250] determine the direction of activity of transport so that the potassium distribution with respect to the cell (higher K+ inside or outside the cell) oscillates.

Fourth, I proposed an *enzyme clock* model based on the facts: (1) pinealectomy results in arrhythmia; (2) the isolated pineal can generate NAT rhythms; (3) the properties of NAT correlate with circadian phenomena (entrainment, freerun, photoperiod, LL attenuation, light and dark pulse PRCs, skeletons);

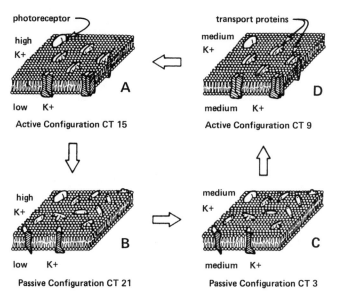

Figure 8.14 Representation of the membrane model for the circadian clock which involves changes in the arrangements and sizes of the membrane transport proteins that are embedded as particles in the lipid bilayer that makes up the membrane. (From Njus, Sulzman, and Hastings, with permission [242]. Reprinted from *Nature*, volume 248, pages 116–120, Copyright © 1974 Macmillan Journals Limited).

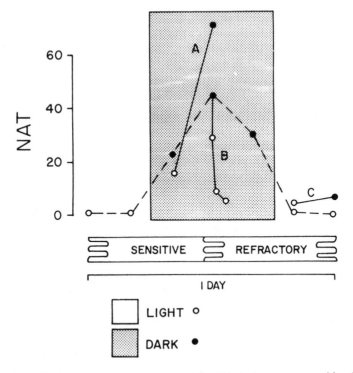

Figure 8.15 An enzyme model for the biological clock was proposed by the author based on the properties of N-acetyltransferase in the eye and pineal.

    NAT exhibits a rhythm in a light-dark cycle (dashed line). Dark cannot stimulate NAT in the light-time (C), but only during the animals' expected dark-time (rectangle, A) based on its prior lighting history. Light rapidly suppresses NAT (B).

    The enzyme clock model has been used to explain photoperiodic time measurement (including light-break and resonance experiments) and circadian rhythm phase shifting by light and dark pulses [33, 35, 38, 43].

and (4) the rapid resetting of the NAT cycle (Fig. 8.15) [35, 41, 43]. In particular, the characteristics of the NAT rhythm can provide an explanation for photoperiodic control of reproduction, and for phase shifting by light and dark pulses. An interesting thing about the enzyme NAT is that, irrespective of the nocturnal or diurnal life style of an organism, NAT is synthesized in the dark-time. In considering the temporal organization of events in the pineal gland and experiments in which the light period prior to a night is manipulated experimentally, the current form of the model includes the suggestion that there is an *initiating process* that recurs each day in the early light-time and that programs the subsequent dark-time events.

    As the reader can see, the models for the circadian clock proposed above are not necessarily mutually exclusive. One could visualize a grand scheme drawing aspects from each of these models to provide a circadian pacemaker.

9

A
TEMPORAL
BESTIARY

"Destruction of the clock by a hypothalamic lesion eliminated all manifestations of the clock in the record of eating
times . . . the rat ate at intervals of 40 to 60 minutes throughout the 24 hours."

Curt Paul Richter, 1967 [283]

"The pineal organ of the house sparrow, *Passer domesticus*,
is essential for persistence of the circadian locomotor
rhythm in constant conditions. Upon removal of the pineal
body, activity becomes arrhythmic . . . Our data demonstrate that the pineal organ is a crucial component of the
endogenous time-measuring system of the sparrow."

Suzanne Gaston and Michael Menaker, 1968 [162]

SOME REAL CLOCKS!

For some organisms, the *black box* has been opened—there is information
concerning their actual circadian pacemakers, the biological clocks. In a way,
it was Pandora's box that was opened when investigators peeked in the black

boxes, because locating the potential oscillators has produced many questions that have led to feverish investigation.

It is my opinion that we now have in our grasp compelling explanations for seasonal photoperiodism and ideas as to the solution of how organisms keep circadian time. It is my belief that the basic circadian pacemakers will have a single escapement universally derived from common anatomical components of cells and biological molecules. I have full faith that the circadian clock will be elegant in its simplicity and ubiquitous among organisms.

For each organism, I will try to review: why the organism has been a popular subject, the methods used to study rhythms, the clock properties (freerun, entrainment, Zeitgebers, resetting), and the anatomical and chemical components (photoreceptors, pacemakers, and slave oscillations).

Table 9.1 lists possible circadian pacemakers and the species in which they are thought to function.

RATS, A CLOCK IN THE HYPOTHALAMUS

Traditionally, rats win the race for favorite laboratory subject. Now, domesticated, usually albino, Norway laboratory rats (*Rattus norvegicus*) are routinely purchased from commercial suppliers, but Richter mentions that "the wild

TABLE 9.1   Some Possible Circadian Pacemakers[1]

Species	Pacemaker
house sparrows [162, 330]	pineal, SCN
Java sparrows [129]	pineal, SCN
pigeons [128]	pineal, eyes
chickens [218, 244]	pineal, eyes
quail [347]	pineal, eyes
lizards [348]	pineal
human [357]	SCN
hamsters [313]	SCN
rats [311]	SCN
ground squirrels [394]	SCN
roaches[245, 292]	optic lobe[2]
crickets [214]	optic lobe
silkmoths [343]	optic lobe
fruit flies [174]	brain
sea hare [212]	eye
African toad [29]	eye

[1]Locations that have been postulated to be circadian pacemakers because ablation abolishes rhythms or because the structures are capable of generating rhythms
[2]supra-oesophageal ganglion of the brain

rats used in this study came from all parts of the city" [281] and noted that "many thousands of captured live wild rats have been studied in our laboratory." In any case, Richter's recordings from rats are among the best showing circadian rhythms in rodents (Fig. 9.1) [282]. Rat locomotor activity is traditionally measured by giving the rats access to running wheels; indeed, one version of the running wheel apparatus has a very large wheel with a very small nest box, which reduces the rat's opportunities to avoid recording.

Rats are nocturnal and they entrained to light-dark cycles. Light is detected by the eyes because Richter's data showed blind rats freerunning with shorter-than-24h periods in the presence of light-dark cycles. It may also be possible to get rats to entrain to periodic food presentations [130]. Rats can be reset by light-dark cycles; Honma published a phase response curve for 30-min light pulses with −2.9h delays at c.t. 15 (early subjective night) and 1.4h advances at c.t. 23 (late subjective night) [179].

Another rhythm whose elucidation has involved rats as subjects is the pineal N-acetyltransferase activity (NAT) rhythm. The NAT rhythm entrains to light-dark cycles, its shape is modified by photoperiod, it persists in DD, it is damped by LL, and it is reset by light pulses [40, 183, 184, 198]. The photoreceptor is the eyes [198]. Although the rhythm of rat pineal NAT was modified by photoperiod [183], the photoperiodic control of reproduction in laboratory rats is less well defined than in some other species (e.g., hamsters) so that rats have not been the organism of choice for elucidating calendar functions.

When Richter obtained arrhythmia in rats following lesions of the hypothalamus [283], his work showed the way to locate a circadian pacemaker in the rat. Stephan and Zucker reported in 1972 that:

"Bilateral electrolytic lesions in the suprachiasmatic nuclei permanently eliminated nocturnal and circadian rhythms in drinking behavior and locomotor activity of albino rats. The generation of 24-hr behavioral rhythms and the entrainment of these rhythms to the light-dark cycle of environmental illumina-

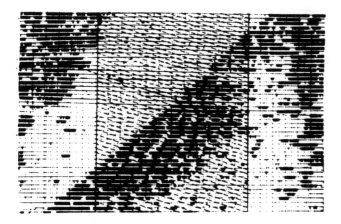

Figure 9.1 Freerunning locomotor activity of a wild Norway rat blinded and kept in LD12 : 12 with dark from 6 p.m. to 6 a.m. (vertical lines). Each horizontal line is 24h of wheel running data, the lines are arranged vertically in chronological order. (After Richter [282].)

tion may be coordinated by neurons in the suprachiasmatic region of the rat brain." [311]

The early work generated interest so that it was followed by a spate of papers showing effects (or lack of effects) of SCN lesions on the circadian rhythms in rats, other rodents, and other mammals (Figs. 9.2, 9.3). Lesioning the SCN of rats also abolished the rhythm of pineal NAT (Fig. 8.10) [235].

With three structures—eyes, hypothalamus, pineal—located for the circadian system in rats, it has been possible using ablation techniques to map out a *hierarchy* for the control of the circadian rhythms of rats (Figs. 8.8, 8.10). The sequence of events is as follows.

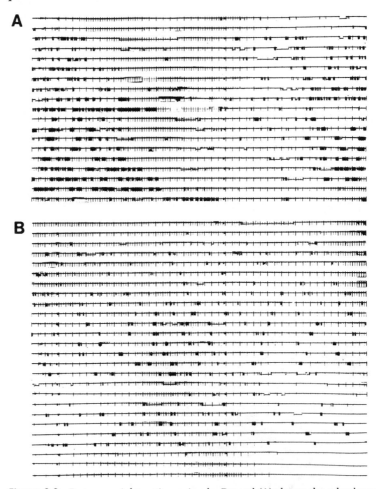

Figure 9.2   Rats are night-active animals. Record (A) shows the wheel-running activity of an albino laboratory rat freerunning in constant dark. Record (B) is the apparently arrhythmic activity of the same rat with lesions of the SCN. Each line is 24h of data. (Riebman and Binkley, 1977.)

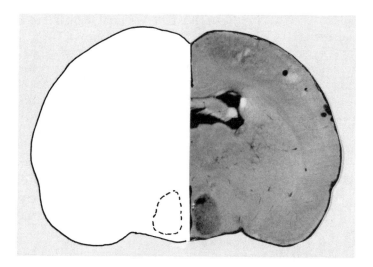

Figure 9.3 The suprachiasmatic nuclei are a biological clock. On the right is a photomicrograph of half of a rodent brain showing the location, in the hypothalamus, of the paired suprachiasmatic nuclei (SCN). (Riebman, Binkley, and Cioffi, 1977.)

1. Light is perceived by the eyes.
2. Lighting information is conveyed from the eyes to the hypothalamus by a special nerve tract distinct from the optic nerves, the retino-hypothalamic tract (RHT).
3. The SCN is the circadian pacemaker.
4. Integrated information from the SCN entrained to an LD cycle is conveyed by nerves to the superior cervical ganglia (SCG) of the sympathetic nervous system.
5. Axons from cell bodies in the SCG convey the information to the pineal gland.
6. The sympathetic terminals of the SCG axons release norepinephrine (NE) in the pineal.
7. NE stimulates beta-adrenergic receptors on pinealocyte cell membranes.
8. Cyclic adenylic acid (cAMP), a *second messenger,* conveys the signal inside the pinealocyte via a *cascade of reactions* to stimulate new synthesis of pineal NAT.
9. NAT controls the daily rhythm of production of melatonin by acetylating serotonin.
10. Melatonin secreted by the pineal into the bloodstream can act as a time signal for other physiological processes (e.g., prolactin release from the pituitary).

The existence of this pathway constitutes the central dogma of circadian rhythm regulation in rats. In the rat, the pineal is viewed as a *neuroendocrine transducer;* that is, it converts a neural signal, which it receives (from the SCG), to an endocrine signal (melatonin), which it emits. The pineal is not considered a *pacemaker* in its own right; the pacemaker is viewed as residing in the SCN, making the pineal NAT a *slave oscillation.* However, in other species (e.g., birds, Chapter 9) the pineal is a pacemaker. The NAT of pineals of adult rats can be stimulated artificially *(in vitro)* with NE or cAMP, but the pineal cannot generate a rhythm on its own. It may be that the pineal of mammals has lost the rhythm-generating ability during evolution because pineals of immature rats do produce one NAT cycle when they are placed in organ cultures [81].

There is further evidence for the pacemaking abilities of the SCN [280, 304]. 2-Deoxyglucose injections, which may indicate brain activity, show that the SCN is most active during the day. Thus, the SCN has a daily rhythm.

HAMSTERS, PHOTOPERIODISM EXPLAINED

Hamsters have been a favorite experimental subject for studying circadian rhythms of locomotor activity and, because of their dramatic responses to photoperiod, they have been a major organism in which photoperiodic regulation of reproduction has been elucidated. Some species of hamsters hibernate so they have also been used in studies of hibernation. In considering scientific papers about hamsters, it is sometimes necessary to distinguish among the separate species of hamsters. The Golden or Syrian hamster, *Mesocricetus auratus,* has served many children as pets and is probably the animal we associate with the word *hamster.* However, a much smaller species, the Siberian or Djungarian hamster, *Phodopus sungorus,* has also been used extensively for studies of circadian rhythms and photoperiodism.

Hamsters' rhythms are readily studied as these nocturnal creatures industriously spin running wheels at night (Fig. 9.4) [111, 112, 266, 267, 268]. They entrain to light-dark cycles, and they freerun in DD (periods of 23.5h to 24.5h) [121]. Constant light had two effects: it lengthened period (e.g., from 24.14h to 24.44h [111], Fig. 2.3), and it produced splitting (Fig. 5.4). Golden hamsters contributed phase-response curves as early as 1964 [121]; they delayed (e.g., −60 min) in response to 10-min light pulses early in the subjective night (just after activity onset), and they advanced (e.g., 140 min) after light pulses in the late subjective night [121]. Golden hamsters were also phase-shifted by dark pulses [76, 144]; dark pulses (e.g., 6h) advanced the rhythm (e.g., 6h) in the early subjective night and delayed the rhythm in the late subjective night (e.g., −4h) [76]. Davis and Menaker argued against the possibility of social Zeitgebers because some pairs of hamsters separated by a wire mesh exhibited separate freeruns despite the opportunity for social exchange of time information [118]. However, Mrosovsky [241] was able to produce phase shifts in freerunning hamsters by providing a social opportunity (30 min of togetherness in an open cylinder) or a fresh cage.

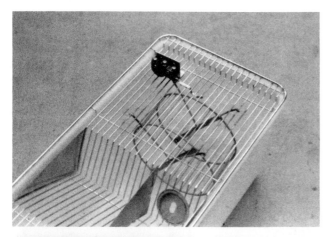

Figure 9.4 Hamsters have been a common laboratory subject for circadian study because their wheel running activity is readily monitored. Here, a golden, or Syrian, hamster is shown with its cage where the wheel has been attached to the cage top to make it easy to clean the cage bottom. Other species of hamsters such as the Siberian hamster have been studied as well.

Pinealectomizing hamsters did not abolish their circadian rhythms; however, lesions of the SCN did disrupt circadian locomotor patterns as well as estrous cyclicity and photoperiodic photosensitivity [313]. Hamsters are believed to possess pretty much the same hierarchy for detection of light signals by the eyes and control of their pineals as rats. One exception appeared to be that hamster pineals were not readily stimulated by adrenergic agents (NE *in vitro* or isoproterenol *in vivo* administered in the subjective day which stimulated rat pineal NAT) [33]. However, NE and isoproterenol stimulated Golden hamster pineals when they were administered in the late subjective night [358, 359].

Hamsters starred in studies of seasonal photoperiodism (Chapter 6) because pinealectomy eliminated the suppression of the reproductive system that occurred when Syrian hamsters were deprived of light (by short photoperiods, DD, or blinding [178]. Initial attempts to alter the duration of nighttime melatonin production in *Mesocricetus* were disappointing with only small changes obtained. However, in the Djungarian hamster, *Phodopus*, the duration of night-time pineal NAT or melatonin was shortened by increased photoperiod [185] and pineal NAT responded to natural photoperiod [310] so that the pineal was capable of transducing photoperiod information into an endocrine signal in the hamster, just as was shown for the chick (Fig. 6.12) [71]. Timed injection (Fig. 6.13) and perfusion experiments have been used to prove that hamster reproduction can respond to modified duration of melatonin [93, 312, 336].

The hamster circadian system has responded to chemical treatments in some experiments. For example, 20 percent D_2O, like LL, lengthened hamster period (e.g., 0.4h) [111]. Triazolam, a short acting benzodiazepine used to treat insomnia, phase shifted (e.g., 100 min advance) Syrian hamster running rhythms [346]. Hamster pineals have not been shown capable of persistent circadian rhythms *in vitro*; however, hamster adrenal glands isolated in culture were shown by one investigator to maintain daily cycles (Fig. 9.5) [5].

SPARROWS, A CLOCK IN THE PINEAL

House sparrows have been used as the signature organism for studies of circadian rhythms in a diurnal organism throughout this book. In this section, I will briefly review the general characteristics of their circadian system. What has been most important about sparrows, however, is that they were one of the first organisms in which the anatomical location of a circadian pacemaker was identified.

Sparrows are day-active and entrain to light-dark cycles; the duration of activity lengthens with the duration of light and bimodal patterns occur with highest incidence in the middle range of photoperiods (Figs. 1.18, 3.1, 3.10, 3.11). Sparrows freerun with a period length about 25h in DD; period shortens in LL (Fig. 2.1). Light is a Zeitgeber for sparrows; however, light also has a direct effect—most sparrows hop if the lights are on even in noncircadian light-dark schedules (Fig. 3.2). Sparrows have extraretinal light perception

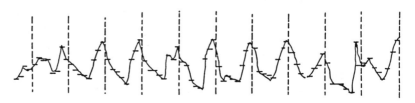

Figure 9.5 Hamster adrenals may be able to maintain rhythms in culture. Here, the rhythm of steroid secretion has been measured for 10 days from a hamster adrenal gland in organ culture. The vertical dashed lines mark off 24h intervals. (After data of Andrews [5].)

[229]; that is, blind sparrows still detect light and can entrain to light-dark schedules (Figs. 3.3, 3.17). Sparrows can also use social cues, probably sound [231], for entrainment (Fig. 3.5), but they only entrain to temperature cycles if the excursion is very large (e.g., 30°F). Constant bright light produces arrhythmia in sparrows (Fig. 5.3). After a time in LL, sparrows placed in DD begin freerunning so that their onset occurs 15h after the LL/DD, as if their circadian clocks were stopped at subjective dusk in LL.

Sparrow phase response curves have been measured in DD with 6h pulses [150] and with 4h pulses during the first 24h of DD following LD cycles (Fig. 4.3) [49, 200]. Advances (e.g., 3.8h) were obtained in the late subjective night, delays (e.g., -1.3h) were obtained in the early subjective night. Four-hour dark pulses imposed on sparrows freerunning in dim LL advanced the rhythm at c.t. 16 (e.g., 2.2h) and delayed it (e.g., -0.7h) at c.t. 0 (Fig. 4.4). Photoperiod modified the shapes of the phase response curves measured for 4h light pulses (Fig. 4.8). The effects of two pulses were additive and could be predicted if it was assumed that the first pulse of a pair phase shifted the response curve immediately so that the second pulse fell upon a shifted rhythm of sensitivity (Fig. 4.10).

Sparrows have seasonal cycles in reproduction (e.g., testis weight, Fig. 6.7) and pineal HIOMT [23]. The reproductive responses can be obtained in the laboratory by altering the photoperiod; the sparrows are able to respond to photoperiod without their eyes [353].

The evidence that the pineal is a biological clock in sparrows began with a study by Gaston and Menaker [162]. Pinealectomized sparrows placed in DD had arrhythmic perching activity and body temperature [30, 50, 162, 230] (Figs. 1.23, 9.6, 9.7). This meant that the pineal was either the site of a circadian pacemaker or involved in the coupling between a pacemaker and the rhythms of temperature and perching. Melatonin injections lowered body temperature and produced roosting behavior in sparrows [32] and continuous melatonin administration produced arrhythmia (Fig. 5.10). Transplanting donor sparrow pineals to the anterior chambers of the eyes of recipient pinealectomized sparrows restored circadian rhythmicity in DD; the timing resumed by the transplant recipient was determined by the phase of the donor sparrow [389]. Disconnecting sparrow pineal stalks *in situ* or exposing sparrows to chemical sympathectomy did not eliminate rhythms [389]. Sparrow pineals exhibit cycles in organ cultures [327], so it seems that the sparrow pineal can function as a circadian pacemaker. Still further evidence of the independence of sparrow pineal oscillations is the fact that sparrow pineals were not stimulated by NE in culture, and sparrow pineals were not stimulated by injected isoproterenol, yet sparrow pineals have the expected nighttime peak in NAT [33, 49]. Lesions of the SCN, however, also abolished rhythms of some sparrows, implying a potential pacemaker role for the hypothalamus [330]. The SCN is a melatonin target [354], so it is possible that the SCN has a *coupler* role.

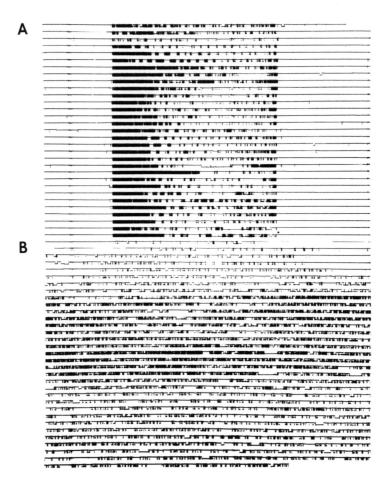

Figure 9.6 Pinealectomy abolished circadian rhythmicity of sparrows and established the pineal as a candidate for a biological clock. The record shown is for a sparrow that was pinealectomized before recording took place. The sparrow entrained to LD12 : 12 with lights-on at 5 a.m. (A). When the sparrow was placed in constant dark (B), its activity was arrhythmic. Each line of the record is 24h of data (Binkley [30], replication of study of Gaston and Menaker [162].)

CHICKENS, A PINEAL CLOCK IN A DISH

Attempts have been made to study locomotor rhythms in diurnal chickens (*Gallus domesticus*). Chicken locomotor activity entrains to LD cycles and the birds have extraretinal light perception [218, 244]. The results for locomotor rhythm studies have sometimes been disappointing because of the relatively diffuse activity patterns of most individual birds. However, well-defined circadian rhythms have been found in chick pineal and retina NAT and melatonin.

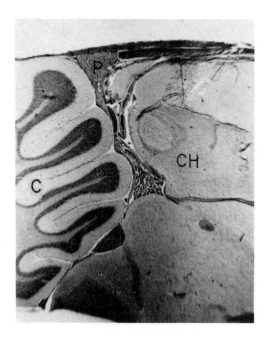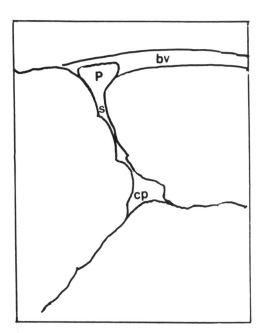

Figure 9.7 The pineal is a biological clock. The sparrow pineal appears in this photomicrograph. The top of the pineal (P) is just beneath the skull, meninges, and blood vessels (bv) and is cone shaped. An elongated stalk (s) extends downward between the cerebellum (C) and the cerebral hemispheres (CH) to the choroid plexus (cp). (Courtesy of Binkley, 1971.)

These rhythms have been studied in whole animals, *in vivo*, by subjecting chicks to light-dark treatments, then killing the chicks at successive time points, and finally measuring NAT or melatonin in the pineal or retina tissues. It has also been possible to study pineal function *in vitro*. Three techniques have been used: (1) organ culture followed by measurement of NAT in many pineals; (2) superfusion culture of single glands with sequential sampling of melatonin in the effluent medium (Fig. 9.8); and (3) culture of dissociated and dispersed pineal cells sampling melatonin in the medium [272, 293, 294, 388].

First, consider the *in vivo* results. Chick pineal NAT has a rhythm in a light-dark cycle (e.g., LD12:12) with a dark-time peak that controls nightly production of melatonin (Fig. 6.9) [62]. The rhythm is a circadian rhythm because it persists in DD with a period near 24h [46]. The rhythm is damped by LL [273]. Both the eyes and extraretinal light receptors play a role in controlling chick pineal NAT, and the pineal has photoreceptor-like structures [273]. Photoperiod modifies the duration and amplitude of night-time NAT production (Fig. 6.12) [71]. It was possible to measure phase response curves for both light and dark pulses. Four-hour light pulses cause advances (e.g., 5h) in the late subjective night and delays in the early subjective night (e.g., − 2h) in the first cycle following the pulses [64]. Four-hour dark pulses cause delays (e.g.,

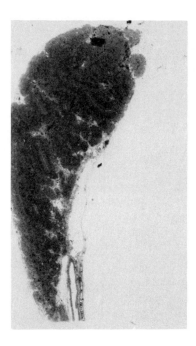

Figure 9.8 The chicken pineal gland has been proved to be a circadian oscillator. The picture is a photomicrograph of the entire pineal gland from a young chicken. (Cioffi and Binkley, 1984.)

– 2h) in the late subjective night and advances (e.g., 4h) in the early subjective night [52].

Second, consider the *in vitro* results. They are particularly important because they constitute evidence (together with the data for sparrows, Chapter 9) that the pineal is a pacemaker. In light-dark cycles, chick pineals or chick pineal cells produce a rhythm of NAT or melatonin formation (Fig. 9.9) [69, 294, 331, 388]. In DD or LL, rhythms of NAT or melatonin formation persist up to 5 cycles but they damp [69, 285, 294, 331, 388]. The pineals are themselves sensitive to light [69]. The pineals carry phase information since the time course of NAT for the first 24h of culture is dependent upon the time the chicks were killed (an early day kill produces a small first cycle, a late day kill produces a large cycle [67, 69].

Thus, the evidence that chick pineals are circadian pacemakers is:

1. Rhythms in NAT and melatonin have characteristics correlated with our expectations for circadian rhythms.
2. Chick pineals and cells can generate rhythms *in vitro* and carry phase information.
3. The rhythms in chick pineals respond immediately (by the first cycle after a perturbing signal such as a light pulse, dark pulse, or reversed phase of the light-dark cycle) [39, 52, 64].

Moreover, it seems likely that individual pineal cells contain the photoreceptor and pacemaker. Chick pineal cells respond to light pulses (e.g., 4h or 6h)—up to

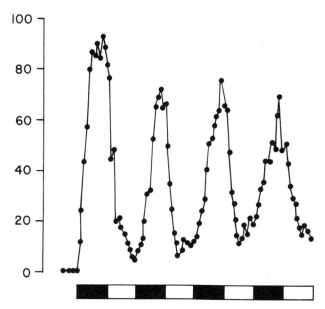

Figure 9.9 The chick pineal exhibits rhythms of N-acetyltransferase and melatonin when it is isolated in cultures. Here, the melatonin rhythm obtained from a single chick pineal is shown; the pineal was maintained in LD12 : 12 for four days of culture. The vertical axis is melatonin (ng/ml); the bar represents the light-dark cycle during the culture and is four days long. (Binkley and Tamarkin, 1980.)

8h delays were obtained in the early subjective night and 8h advances were obtained in the late subjective night [294, 388]. Chick pineal cells also respond to dark pulses with advances in the late subjective day (e.g., 6h) and delays near subjective dawn (e.g., −6h) [388].

However, pinealectomy did not eliminate rhythms in all operated chicks [218]. A possible resolution of this problem has been found in the discovery that chick retinas are also capable of rhythmic melatonin synthesis in much the same manner as the pineals [47, 66]. The presence of rhythmic melatonin generating systems in both the eyes and pineals of some species (e.g., chickens [66], quail [44], sparrows [66], hamsters [66], rats [66], *Xenopus* [28, 29]) seems redundant but may explain experiments where pinealectomy was ineffective [29, 42].

Because of the established pathway (Chapter 9) from the eyes to the pineal for rodents, investigators have examined the possibility of this pathway in birds. Adrenergic agents (e.g., NE, isoproterenol) inhibit chick pineal and ocular NAT and melatonin *in vivo* and *in vitro* [33, 66, 68, 272, 285], in contrast to rats where these agents are stimulatory. Thus, inhibitory adrenergic regulation, presence of extraretinal and pineal light perception, and ability to generate rhythms are all characteristics of chick pineals that differ from the dogma worked out for pineal regulation in rats. However, chick pineal NAT is stimulated with theophylline or cyclic nucleotide analogs [285, 365] and there is a rhythm of cGMP content in the chick pineal *in vitro* [364].

A further pharmacological direction that has been explored using chick pineal tissue has been to show by injecting the protein inhibitor, cycloheximide, that new protein synthesis is necessary for the rise in and continued night-time high NAT [68]. The enzyme, NAT, is very unstable at 37°C; but the activity of NAT can be protected with cold (e.g., 0°C) or affected with compounds related to sulfur metabolism (e.g., protected with cysteamine and acetylcoenzyme A; inhibited with cystamine, N-ethylmaleimide, and taurine) [70].

ROACHES, A CLOCK IN THE OPTIC LOBE

Cockroaches have been a popular insect for circadian rhythm studies because individual cockroach wheel-running activity can be recorded, just as wheel-running can be recorded in rodents (see Fig. 1.7). The roaches that have been studied most are nocturnal (*Leucophaea maderae Fabricus, Bursotria fumigata Guerin, Periplaneta americana Linneas*), although the Asian roach is supposed to be active in and attracted to light. In the laboratory, roaches are cultured in aquaria with wood chips and fed lab chow and water.

The freerunning rhythm of activity lengthened in constant light and was temperature compensated in the range 20–30°C [287]. Interestingly, roaches have two activity components and can be made to shift the component that is dominant by changes from DD to LL (Fig. 9.10). Roaches entrained to 24h cycles begin their activity at lights-out in light-dark cycles and at the temperature maximum in a temperature cycle [288]. Twelve-hour light pulses or 12h cold pulses phase shifted roach rhythms [288]. A PRC was obtained for *Leuco-*

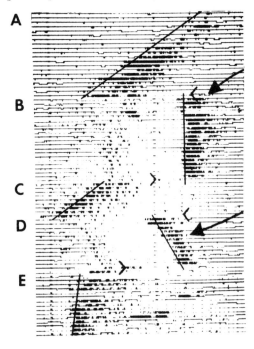

Figure 9.10 Roaches are nocturnal and have well-defined circadian rhythms. Sometimes there are two components. Here, a record from a roach shows the changing of its period length and shifting the bulk of its activity to a second component in response to changes from DD (A,C,E) to LL (arrows B,D). (Courtesy of S. K. Roberts.)

phaea with 6h light pulses with peak delays (− 4h) in the early subjective night (c.t. 14) and peak advances (4h) in the subjective night (c.t. 18) [247].

Roaches have two sets of photoreceptors—the compound eyes and the ocelli; covering the compound eyes blocked entrainment to light but removing the ocelli had no effect [289]. So the eyes are believed to be the photoreceptors for the circadian system of roaches. The site of the circadian oscillator has also been sought successfully in the cockroach. Bisecting the brain produced arrhythmia in *Periplaneta americana*, indicating a neural location for the oscillator [290]. Uwo and Pittendrigh more precisely localized the oscillator(s) in the optic lobes by making lesions at various levels of the nervous system [241b]. Subsequently, Roberts further localized the specific area of the potential oscillator within an optic lobe by the lesion technique [292] (Figs. 9.11, 9.12). Further evidence that the optic lobes were the site of the pacemaker was provided when Page delayed the phase of rhythms of *Leucophaea maderae* by application of cold temperature (7.5°C, 6h) to optic lobes of roaches [246]. Additional evidence was found when optic lobe transplants restored rhythms in 4–8 weeks to *Leucophaea* that had previously had their own optic lobes surgically removed [245]. However, lobectomized roaches can still have exogenously driven cycles because temperature cycles produced rhythms in *Leucophaea* whose optic lobes had been removed [246].

It is possible to relate the biochemistry of roaches that do not have pineal glands (invertebrates) to the biochemistry of indoles in the pineal (vertebrates). Serotonin injections produced a PRC in roaches similar to that produced by light pulses [247]. Moreover, antibodies to the pineal enzyme HIOMT

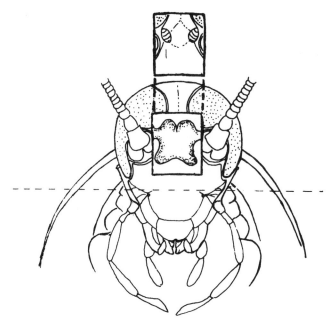

Figure 9.11 A head-on diagram of the cockroach showing (in the rectangle) a view of its brain. (Courtesy of S. K. Roberts.)

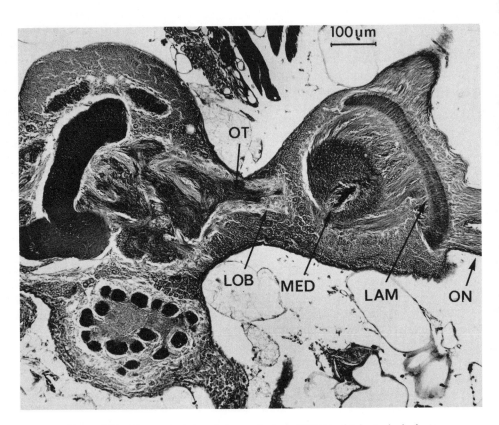

Figure 9.12 The optic lobe of the cockroach (LOB), a biological clock, is shown in a photomicrograph. The structures to the right lead to the roach's eye. The optic tract (OT) carries light information to the roach's brain. (Courtesy of S. K. Roberts.)

bind to roach cells, constituting immunological evidence that roaches have an enzyme characteristic of pineal [335]. Cockroach brains and/or optic lobes had NAT activity (around 20 nmoles/brain/h, comparable to dark-time chick pineal levels) and radioassayable melatonin activity (Fig. 9.13). Thus, while the story for roach indole metabolism is incomplete, there is evidence for involvement of elements of the metabolism characteristic of pineal and retina rhythms—serotonin, NAT, HIOMT, and melatonin. The idea that a common system is involved in rhythm regulation in organisms as disparate as roaches and chickens has to be considered.

SEA HARES, A CLOCK IN THE EYE

Insects are not the only invertebrate group whose circadian rhythms have been studied. The mollusks have also been popular among rhythmologists,

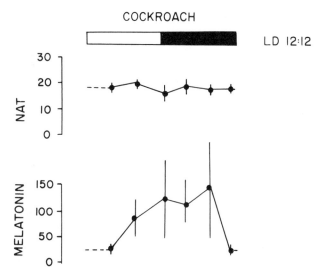

Figure 9.13 NAT and melatonin in the brains of cockroaches. (Roberts, Rollag, Mosher, and Binkley, 1986.)

especially the nudibranch sea hare (*Aplysia californicus*), a marine gastropod (*Bulla bulla*), and the pulmonate slug (*Limax pseudoflavus Evans*). These unlikely organisms were chosen as subjects for special reasons. The nervous systems of mollusks contain giant neurons (e.g., nerve cells a millimeter in diameter); the large neurons permit experimentation on individual cells with penetrating micropipettes and recording electrodes. Moreover, there are relatively small numbers of neurons in these animals so that it is possible to describe the detailed cellular anatomy of their nervous systems. Rhythms in locomotion have been measured in wheels, by television monitoring, and with trippers suspended around the edges of tanks [189, 212, 317].

" . . . individual animals were confined in 20 l polyethylene containers. The sides and bottom of the containers were perforated with 9 mm holes to allow circulation of water, and the container was suspended in a . . . tank to a depth which allowed its mouth to protrude about 6 cm above the water. In early versions we attached Plexiglas rods to the lip of the container so that the rods hung vertically about 2 cm from the inner wall of the container. Each rod carried a mercury switch about 10 cm above the water. As an *Aplysia* crawled around the inner walls, it encountered the rods, deflected them and caused a closure of the mercury switch. The switch closures activated an event recorder. In later versions we replaced the rods and mercury switches with nylon fishing line and gentle action microswitches. In this version, the animal deflected the lines and tripped switches to which they were attached." [212]

It was also possible to record circadian rhythms of electrical activity in the eyes [73, 188, 316].

Aplysia locomotor activity is diurnal, it freeruns, and it entrains to light-dark cycles [191]. The eyes are candidates for circadian pacemaking, but eyeless *Aplysia* entrain [295]. Interestingly, eyeless *Bulla*, which are normally nocturnal, become diurnal in a light-dark cycle. The eyes of *Aplysia* (which appear as barely visible, 0.7 mm dark dots on the animals) are the photoreceptors. In his review article, Jacklet described the locomotion:

> "Locomotion in *Aplysia* is performed by undulatory stepping of the foot. The anterior part of the foot is detached from the substrate, extended, and reattached. Then, waves in the foot pull the animal forward, and the posterior foot is released and advanced to a new position." [191]

As mentioned, however, it was the properties of the nervous system that captured the interest of investigators. In 1965, Strumwasser recorded a cycle in the spike output of a cell in the parieto-visceral (abdominal) ganglion [316]. In 1969, Jacklet recorded circadian rhythms in neural output from *Aplysia* eyes that had been isolated *in vitro* (Fig. 9.14) [188]. The isolated eyes, then, contain both photoreceptor and pacemaker function [73, 190, 191]. The rhythm of eyes freeran (tau = 23.9h to 27h), shortened in LL, was phase shifted by light pulses, and was temperature compensated. Full resetting properties may require attachment of the eyes to the cerebral ganglion. Activity from the eyes was highest in the subjective day.

The ability to study the eyes *in vitro* means that it is possible to experimentally manipulate the chemical milieu to attempt to determine the basis for the photoreceptive and pacemaking functions by examining the effects of various compounds upon the phase, amplitude, and period length of the rhythm measured from the optic nerve. The involvement of membranes has been supported by changes wrought with ions (potassium, lithium, manganese, valinomycin, ethanol, lanthanium, nystatin). Yet other changes have been produced with inhibitors of protein synthesis (cycloheximide, anisomycin, puromycin). Because the *Bulla* eye contains relatively few cells, Block and co-workers [74] were able to develop a model for the *Bulla* retina in which the basal retinal neurons are "entraining photoreceptors as well as candidate circadian pacemakers" [74, 75].

Vis-a-vis my discussion of the potential role of indoles in pineal rhythms, it may be noteworthy that serotonin is present in the *Aplysia* eye (50 ng/mg protein) and that it shifted the phase of the *Aplysia* eye rhythm [106, 152]. Serotonin delayed the rhythm at c.t. 21−5 and advanced the rhythm at c.t. 5−14. This is 180 degrees out of phase with the light pulse PRC measured for *Aplysia*, but I suggest that it may correlate well with the PRC for dark pulses (not measured for *Aplysia* but discussed for other organisms). However, using 8-bromo cGMP (a cyclic nucleotide analogue), Eskin and co-workers did mimic the light pulse PRC and suggested cyclic nucleotide involvement in the photoreception process represented by light-induced phase shifts [153].

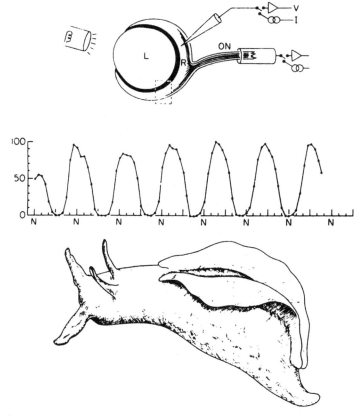

Figure 9.14 The sea slug, *Aplysia*, represents the mollusks whose accessible nervous systems have made them a subject of investigation. In particular, the eyes have been studied. (With permission, Palmer [249], and Jacklet [191].)

GONYAULAX, A CLOCK IN A CELL

Investigators studying multicellular organisms, such as *Aplysia* or hamsters, worry about and investigate the interactions of oscillators and the interactions of cells within a pacemaker. But there is good evidence that the circadian clock can be contained within single cells, and the bulk of that evidence comes from investigations of single-celled organisms. The single-celled organisms are usually studied as populations. One of the single-celled organisms that has received considerable attention is the organism of the *red tide*, a marine dinoflagellate named *Gonyaulax polyedra*, which can reach high concentrations in sea water (Fig. 9.15) [25]. This organism contains chlorophyll and therefore Sweeney includes it among the plants [320, 325]. The organism has been useful for circadian rhythm studies because it is capable of *bioluminescence* as described by Sweeney and Hastings:

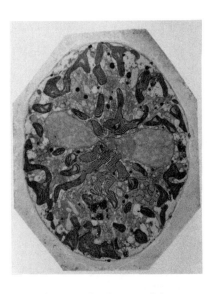

Figure 9.15 A longitudinal section of *Gonyaulax polyedra*. (With permission, Sweeney [320].) The picture is the frontispiece in Dr. Sweeney's book and is credited by Academic Press to Dr. G. Benjamin Bouck.

"The luminescence of cultures of the marine dinoflagellate, *Gonyaulax polyedra*, receiving light from the window, is dim during the day and about sixty times brighter at night. These variations may be described as a type of inherent diurnal rhythmicity, since a number of experiments demonstrate that the periodicity is maintained in the absence of normal day-night variations in light intensity. If the cultures are transferred from the window to total darkness at any time, the rhythmic changes in luminescence continue to occur, approximately in phase with cells left in the window. Continuous exposure to light will inhibit the rhythm, but it may be initiated in such light-treated cultures at any time by placing them in total darkness. The rhythm will then persist for about four days. The amount of luminescence which develops at night is directly related to the amount of light received during the day. Light will also directly inhibit luminescence at any time during the night. The changes in luminescence are at least partially due to synthesis and destruction of luciferin and luciferase." [325]

Disturbing the seawater of the flasks in which populations of the cells swim (e.g., with bubbles or acetic acid) causes them to emit flashes of light, which can be measured with a photomultiplier photometer. The amount of bioluminescence is low during the day, but high at night, and it persists in LL with a circadian rhythm (Fig. 5.9). Bioluminescence is not the only rhythmic parameter that has been measured in *Gonyaulax*—rhythms of oxygen production, carbon dioxide fixation, cell division, luciferin activity, luciferase activity, luciferin-binding protein, chloroplast structure, chloroplast position, membrane particles, protein synthesis, and activity have also been measured. The luciferin measures are biochemical, associated with the production of bioluminescence.

As noted in the quotation from Sweeney and Hastings above, *Gonyaulax* rhythms damp out in DD, which is probably due to the need for light for photosynthesis, but the circadian rhythms freerun in LL. Bright light damps the rhythm in cultures of *Gonyaulax,* but it takes a long time for the LL to produce a state at which LL/DD resets to c.t. 12 [324]. Light acts as a Zeitgeber, and the *Gonyaulax* rhythm is temperature compensated. The circadian rhythm is reset by 3h light pulses (– 10h delays at c.t. 15, + 12h advances at c.t. 18) [322]. Roenneberg and Hastings [295] looked at the effect of light intensity upon tau using red, yellow, white, and blue light. Tau increased (e.g., from 23.5h to 27h) with increasing light intensity in red and yellow light; tau decreased (e.g., from 23.5h to 22h) with increasing light intensity in white and blue light. The investigators argue that this is evidence for two photopigments involved in photoreception for the circadian rhythm of *Gonyaulax.*

Since *Gonyaulax* is a single-celled organism, it seemed likely that single cells contained all the machinery necessary to generate circadian rhythms and to respond to light. Single *Gonyaulax* displayed rhythms of oxygen production [319]. Still, Broda and others [83] concluded that there was *circadian communication* (social synchronization?) between single *Gonyaulax* cells from results obtained by mixing two populations of *Gonyaulax* that were 11h out of phase, but the conclusion has not been supported by studies of other investigators (Sulzman and coworkers [318], and Sweeney, personal communication).

Biochemical measurements were made using cultures of *Gonyaulax.* "Violent fluctuations in connection with clock activity" [175] were not found in DNA, but there was a rhythm in RNA [367]. The activity of the enzyme, luciferase, peaked at mid-dark, and there was also a peak in luciferin, the substrate for luciferase [225]. *Gonyaulax,* bathed in aqueous medium, is amenable to biochemical studies because compounds can easily be added to the medium and effects on the period and phase of the circadian rhythms can be measured. Using this approach, a variety of compounds have been tested. Ethanol shortened [338] and D_2O lengthened the period of *Gonyaulax* rhythms [223].

The potential anatomical-chemical basis for the circadian pacemaker in *Gonyaulax* has been considered [322]. The model building has taken three directions. First, a chronon type model based on transcription was ruled out because actinomycin D did not change the period and there was a lack of fluctuations in DNA. But the RNA rhythm argues for a role of transcription [367]. Second, models involving protein synthesis, the process of translation, have been better supported because protein synthesis inhibitors (e.g., cycloheximide [126, 279], anisomycin [337]) shifted the phase of *Gonyaulax* rhythms, and protein synthesis (e.g., leucine incorporation) has a circadian rhythm. Third, membrane hypotheses (Fig. 8.14) [242, 243, 322] have been supported by the lengthening of tau by ethanol, the temperature independence of tau, the fact that phase shifts can be obtained with valinomycin, and detection of daily changes in cellular potassium and membrane structure [323]. Fourth, in describing the response of the *Gonyaulax* rhythm of bioluminescence, I am struck by the parallels to the properties of pineal NAT (especially the dark-time plummet in response to light and the effect of prior light on amplitude of the rhythm) so that I have wondered whether there is a class of enzymes capable of similar circadian responses and involved in the circadian clockworks.

EUGLENA, ANOTHER CLOCK IN A CELL

Another single-celled organism in which circadian rhythms have been studied extensively is the algal flagellate, *Euglena gracilis* (Fig. 9.16). It is possible to measure rhythms in cell division, mobility, morphology, photosynthetic capacity, metabolic parameters, amino acid incorporation, enzyme activities, and cell susceptibility (e.g., to alcohol) [135, 136, 138, 208]. Edmunds describes automated measurement of cell division cycles in *Euglena* by measuring numbers of cells:

> "*Euglena* . . . was grown . . . in 3-liter serum bottles on a modified . . . inorganic medium at 25°C . . . The cultures were magnetically stirred (heat compensated) and aerated with 575 to 625 ml of air per minute while being subjected to alternating periods of light (14 hr) and dark (10 hr) by a clock timer. Illumination (3500 lux; saturating) was furnished by a bank of three . . . 40-watt cool white fluorescent bulbs. The cell number was automatically monitored every 2 hours with a(n) . . . automatic pipetting machine, a miniaturized fraction collector, and a cell . . . counter." [132]

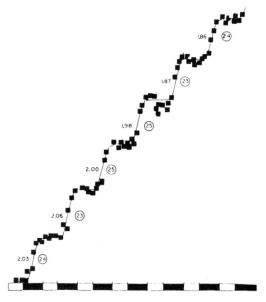

Figure 9.16 The cell division rhythm of *Euglena* is entrained by a light-dark cycle. The horizontal axis represents the LD10 : 14 light-dark cycle to which cultures of *Euglena* were exposed; the vertical axis represents number of cells. The small numbers to the left of the graph indicate step sizes (ratio of number of cells/ml following a division burst). The period (tau) (hours between successive onsets of cell number increase) for each step is given in the small numbers to the right of the graph. (After Edmunds [135]).

Using this system, Edmunds has been able to establish circadian rhythm phenomena in *Euglena* cell division cycles [134]. *Euglena* divide in the dark-time and the cycle entrains to 24h cycles (e.g., LD 4:20, LD8:16, LD10:14, LD14:10). The limits of entrainment are 20–28h, and the rhythm entrains to skeleton cycles. The freerun persists for eight days in a typical experiment in LL (800 lux, 25°C) with a period of 24.2h, under short LD cycles (LD2:6, LD1:3, LD1/4:1/2), and under random lighting regimens. The rhythm exhibited temperature compensation [4]. The rhythm was lost in bright LL (3500–7000 lux) or in DD [133]. Rhythms were initiated in asynchronous cultures by a *step-down* of light intensity (an L/D). Using experiments in which out-of-phase cultures were mixed Edmunds concluded that the cells do not communicate time information. A PRC for light pulses (3h, 7500 lux, 25°C) had peak advances (+ 10h) at c.t. 22 and peak delays (− 10h) at c.t. 20, that is in the dark-time. Dark pulses (3h) also produced phase shifts, mostly delays (e.g., 8h or less) that were out of phase (180 degrees) with the shifts produced by light pulses [220]. There was a singularity; a 3h 40–400 lux pulse at c.t. 23.1–0.4 caused cultures to become arrhythmic [139, 220]. Investigators using *Euglena* have suggested a phytochrome, or perhaps a blue-light photoreceptor [164] for *Euglena*.

Edmunds has generalized his findings in *Euglena* to make models of circadian clocks based on the cell cycle (Chapter 8). Edmunds and Laval-Martin tabulated 17 unicellular systems for which cell division rhythms have been documented. Most of the division occurs, as in *Euglena*, in the subjective night. The scientists point out that "the CT at which all delay or maximum advance phase shifts are achieved corresponds to the approximate position of the (cell division cycle) during which division occurs (commencing about CT12) and to the first few hours of the G_1 phase" [138]. The correlations of circadian changes with events of the cell division cycle have been the basis for the invention of models for *cell cycle clocks*. The idea has been hard to cell (sic) to other investigators, e.g., D. A. Gilbert, who expressed the skepticism poetically:

> "The living cell is most surely dynamic,
> Its behavior so very erratic;
> Why is it so true
> That all but a few
> Insist on treating it static?
>
> There is an awfully big schism
> Between those proposing a rhythm
> For the cell as it grows
> And all of their foes
> Who just won't agree with 'em." [163]

As for *Gonyaulax*, the *Euglena* rhythms are susceptible to manipulation by chemical modification in the medium. Based on the modifications obtained, the researchers have suggested various gears in the cell cycle clock. For example, a closed loop model for a self-sustaining oscillator [164] proposed consists of NAD+ or a photoreceptor causing Ca^{2+} to increase in cytoplasm (by efflux from the mitochondria or influx across the plasmalemma); the calcium forming an activated Ca^{2+}-calmodulin complex; the complex then causing activation of NAD+ kinase and inhibition of NADP+ phosphatase so as to decrease the rate of NAD+ production. However, the investigators note that

> "Our model does not explain either the long period-length or the steady-state temperature compensation characteristic of circadian rhythmicity, primarily because of a lack of hard data . . . We have proposed a model—not for the clock— but for one oscillator in what is most probably a cellular clockshop." [164]

In another example of chemical manipulations, Edmunds and co-workers [137] found that some mutants of *Euglena* lost the ability first to freerun then to entrain to LD cycles, but that disulfide and sulfhydryl-containing compounds (cysteine, methionine, mercaptoacetic acid, dithiothreital, sodium thiosulfate, sodium monosulfide) improved synchrony and reversed their losses. Based on the fact that rhythms were restored at the phase expected if the underlying oscillator had been running throughout the experiment, rather than a phase dependent upon the chemicals, the investigators favored a hypothesis that the agents were acting on the coupling between the underlying clock and the measured cell division rhythm.

FLIES, A BRAIN CLOCK ON A CHROMOSOME

Flies have always been a popular subject for geneticists because of their short generation time, ease of rearing in captivity, and economic importance. Circadian rhythm scientists have also used flies of many species as subjects [302]— flesh flies (*Sarcophaga argyrostoma*), flies of the *Drosophila* genus (*melanogaster, phalarata, littoralis, victoria, pseudoobscura*). The fruit fly (*Drosophila pseudoobscura*) has held its place with scientists as the laboratory rat of the insect world. Rhythms of locomotor activity have been measured in individual flies by monitoring their movements with infrared light [204]. Flies begin life in an egg, which hatches into a larva, a maggot. The maggots enter a period of arrest of growth and development called diapause (sometimes to enable them to survive winter), which is frequently induced by photoperiod. To attain adulthood, a fly sometimes forms a pupa during diapause in which it resides in a case and undergoes metamorphosis from the larval form to the adult form. The adult fly, winged and sexually mature, emerges from the pupal case, and the process of emergence is called eclosion. Eclosion is rhythmic; in the feral fruit fly it occurs in the early morning. Originally, investigators remained in the laboratory around the clock to record fly emergences. Zimmerman and Pittendrigh describe one of the apparatuses used to measure the eclosion rhythm of fruit flies:

> "Automatic devices, incorporating a rigorously controlled light and temperature environment, have been developed in the Princeton University laboratory for the assay of the time of emergence of adult drosophilids from populations of pupae . . . Pupae are reared in plastic boxes in an LD12:12 light/dark cycle at constant temperature (20°C or 28°C). They are harvested by a flotation method and glued to a small brass holding plate. The pupae on the holding plate are enclosed by a Lucite cover with a tapered base that vents above a vial of detergent solution. The mounting plate is suspended from a solenoid which is activated every 30 min, lifting and dropping (against a rubber stop) the system and thus shaking out, into the vial of detergent, all flies which have emerged in the previous 30 min interval. The vials of detergent are assembled in a circular tray which is rotated—1 vial/hr—by a spring loaded escapement mechanism. The trays of twenty-four vials are changed daily, and the counts of flies in the hourly collection yield an assay of the eclosion rhythm." [392]

Thus, the number of fruit flies emerging at a given hour is obtained as dead flies/vial-h data, which can be plotted as a time series (Fig. 9.17).

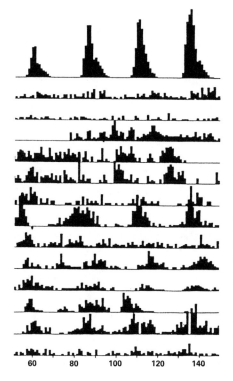

Figure 9.17 Fruit flies have rhythms of eclosion as well as locomotor activity. Here, *Drosophila* eclosion patterns are shown for hours 50–100 after transfer to constant dark. The vertical bars represent number of flies. The top histogram represents control cultures. The 13 cultures below were treated with light pulses with different irradiances (0.01 to 12,500 microwatts/cm²) and durations (50,000 sec to 0.04 sec) of light which attenuated the rhythm to varying degrees. (After Chandrashekaran and Engelmann [100].)

Measuring the eclosion rhythm of a population of flies obscures individual differences but has the advantage that large numbers can be used. In making conclusions from eclosion rhythms, then, a reservation is that circadian oscillators (plural), and not a single pacemaker, are being examined. The locomotor method can capture rhythm information for single flies, but the numbers of flies that can be monitored is limited.

The eclosion rhythm of *Drosophila* entrains to LD12:12 [87, 90] and persists in constant dark (e.g., for six cycles; the number of cycles that can be measured is limited by the flies' life cycle) [87]. It is temperature compensated [51]. Cycles among flies of a population can be synchronized with an LL/DD transition or a light pulse (e.g., 4h). Following a final photoperiod of more than 12h, lights-off acts as a time cue so that emergences recur at a time that extrapolates to 15h after the L/D (Fig. 5.6) [259]. PRCs for light pulses of different intensities, color, and duration have been reported for *Drosophila* [90, 95, 96, 98, 99, 256, 258, 383]. It is possible to obtain phase shifts in *Drosophila* with 1/2000 sec flashes of light [256].

Drosophila rhythms also can follow temperature cycles—emergence follows the low point of a temperature cycle as long as that low point occurs from the late subjective night to the late subjective day. During most of the subjective night, the flies do not follow the temperature minimum, but instead emerge near dawn. The range of phases at which the flies tend not to emerge (in the early night) is called the *zone of forbidden phase*. Locomotor activity of individual flies freeran with a period of about 25h [204]. Action spectra for phase shifts point to a photopigment that is sensitive in the blue region and that may not be a carotenoid [159, 310].

So called *formal properties* of the circadian biological clock have been studied with flies. First, using protocols with two pulses of light [95, 260], investigators concluded that *Drosophila* are reset by the second pulse in a manner that was predicted by assuming that the first pulse completely reset the resetting sensitivity rhythm (Chapter 4). Because of this result, they concluded that light pulses produce an *instantaneous* resetting of the underlying circadian pacemaker. Second, the *singularity* was discovered by examining the eclosion rhythm in *Drosophila* (Chapter 5) [383]. Third, in experiments which should be repeated, Chandrashekaran and co-workers [99] evaluated the roles of dawn and dusk (Fig. 5.7) and concluded that "light pulses shift phase with the off transition during the first half of the night (dusk effect) and shift phase with the on transition during the second half of the night (dawn effect)" [99]. Fourth, enduring models such as the *theory of entrainment* and the presence or absence of A and B oscillators (Chapters 3 and 4) [97, 256, 258] have been based on the properties of fly rhythms and fly PRCs. In the A and B oscillator model, Pittendrigh [256] proposed that there is a light-sensitive pacemaker, the A oscillator, that is immediately reset by light. To explain the synchronization to temperature, he proposes a temperature-sensitive B oscillator. He postulated a *zone of forbidden phase* (occurring at night) to account for the fact that the rhythm only follows the temperature minimum through the subjective day.

Konopka and Benzer [204] used *Drosophila melanogaster* to isolate mutant flies in which the eclosion and locomotor rhythm characteristics were different from the normal near-24h rhythm. They subjected flies to a chemical that caused mutations (ethyl methane sulfonate), examined 2000 male flies with abnormal emergences, and made genetic maps of three rhythm mutants. One mutant was arrhythmic, a second had an abnormally short freerun (tau = 19h), and a third had an abnormally long freerun (tau = 28h). The investigators suggested that the mutations involved the increasing, decreasing, or elimination of activity of one gene on the flies' X chromosome, the *per* locus [24, 204, 308]. The genetic region is also important for the high-frequency wing vibrations of courting males [171].

To discover the anatomy of the circadian pacemaking system in *Drosophila*, Handler and Konopka [174] transplanted brains from flies with the short period mutation, *per^s*, to the abdomens of the arrhythmic mutant, *per^o*. Four of 55 flies receiving transplants exhibited at least three cycles of short-period activity rhythms. The investigators concluded that the brain carried the pacemaker and argued for a humoral diffusible factor secreted from the donor brain. The small size of fruit flies makes it difficult to isolate photoreceptors, but the work that has been done has permitted investigators to argue for nonvisual light reception, to suggest the lack of use of carotenoid pigments, and to propose that light is absorbed by the brain for circadian rhythm control [302, 391].

Melatonin may play some role in fruit fly circadian rhythms. Homogenates of *Drosophila* can produce melatonin if provided with radioactive substrates (5-hydroxytryptophan, serotonin, or N-acetylserotonin) and melatonin injections reduced mating speed and oviposition rate in females [156].

Although some insects appear to use hourglasses or nonrepetitive interval timers for photoperiodic events (such as the control of entry into pupal diapause), other insects, such as the flesh fly, have photoperiodic time measurement that is dependent upon circadian systems [302].

10

WINDING DOWN
WITH
APPLICATIONS

"Early to bed and early to rise, makes a man healthy, wealthy, and wise."

Benjamin Franklin [160]
Poor Richard's Almanac

"If a man does not keep pace with his companions, perhaps it is because he hears a different drummer. Let him step to the music which he hears, however measured or far away."

Henry David Thoreau [340]
Walden

"This *(shock-phase)* hypothesis consists of the following postulates:
1. That each one of the functioning units of the organism—cells, follicles, glomeruli, neurons, etc.—has an inherent cycle, the length of which is characteristic of that organ.
2. That these units, although physically closely bound together, may operate independently of one another.

Thus, they may all be active, or inactive, or at various stages between activity and rest.

3. That in a normal healthy organ these units function out of phase, that is, some are active while others are resting or at various intermediate stages, thus insuring a fairly constant level of productivity.

4. That under certain circumstances these randomly functioning units of an organ may all be synchronized by a shock or trauma, or by other forms of interference, thus revealing the lengths of the inherent cycles of the individual units.

5. That under certain circumstances the synchronized units of an organ may be desynchronized by various forms of interference, thus restoring the organ to its normal non-cyclic productivity."

Curt Paul Richter [281]

APPLYING CIRCADIAN RHYTHM PRINCIPLES

The application of circadian principles to animal husbandry is already in use. For example, it is possible to control the timing of fertility of economically important species by manipulation of the photoperiod. For example, it is possible to advance the breeding activity of ewes by 10 weeks using either short photoperiod or melatonin [196, 271]. As another example, chickens are induced to increase egg production and growth with constant light. Planning for seasonal changes in weather and light is necessary for success in raising most agricultural crops. If the reader is a home gardener, he has made first-hand use of planting guides to time his vegetables and flowers.

Probably of more interest to the reader are the possible applications of circadian rhythm principles to human endeavors. We know how circadian systems work, and we can make intelligent suggestions for coordinating an individual's activities with the timing of the physiology of his body.

ENTRAINMENT BY ALARM CLOCKS?

We take for granted the organization of our time into a waking day and a sleeping night. This makes us a diurnal species. Physicians, psychologists, and scientists have documented a host of human physiological measurements that

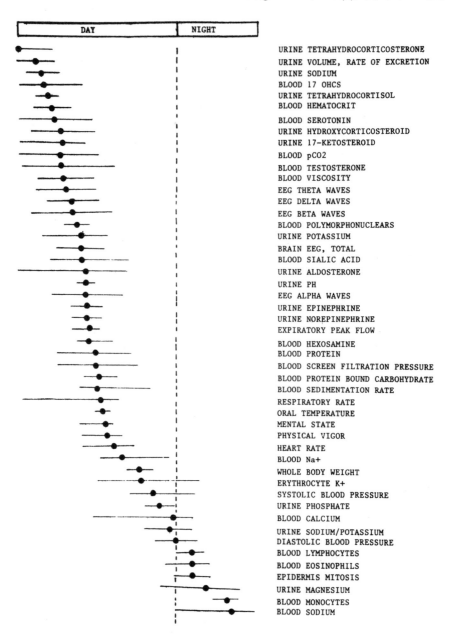

DAY	NIGHT

URINE TETRAHYDROCORTICOSTERONE
URINE VOLUME, RATE OF EXCRETION
URINE SODIUM
BLOOD 17 OHCS
URINE TETRAHYDROCORTISOL
BLOOD HEMATOCRIT

BLOOD SEROTONIN
URINE HYDROXYCORTICOSTEROID
URINE 17-KETOSTEROID
BLOOD pCO2
BLOOD TESTOSTERONE
BLOOD VISCOSITY
EEG THETA WAVES
EEG DELTA WAVES
EEG BETA WAVES
BLOOD POLYMORPHONUCLEARS
URINE POTASSIUM
BRAIN EEG, TOTAL
BLOOD SIALIC ACID
URINE ALDOSTERONE
URINE PH
EEG ALPHA WAVES
URINE EPINEPHRINE
URINE NOREPINEPHRINE
EXPIRATORY PEAK FLOW
BLOOD HEXOSAMINE
BLOOD PROTEIN
BLOOD SCREEN FILTRATION PRESSURE
BLOOD PROTEIN BOUND CARBOHYDRATE
BLOOD SEDIMENTATION RATE
RESPIRATORY RATE
ORAL TEMPERATURE
MENTAL STATE
PHYSICAL VIGOR
HEART RATE
BLOOD Na+
WHOLE BODY WEIGHT
ERYTHROCYTE K+
SYSTOLIC BLOOD PRESSURE
URINE PHOSPHATE
BLOOD CALCIUM
URINE SODIUM/POTASSIUM
DIASTOLIC BLOOD PRESSURE
BLOOD LYMPHOCYTES
BLOOD EOSINOPHILS
EPIDERMIS MITOSIS
URINE MAGNESIUM
BLOOD MONOCYTES
BLOOD SODIUM

Figure 10.1 Acrophase charts for human physiological measurements and susceptibility beginning in the morning at the left and ending at the end of the night on the right. A point represents the time of the peak as it occurs over 24h; horizontal lines are a measure of variation; approximate times of day and night are indicated at the top; the horizontal axis is 24h. (After data in Luce [215].)

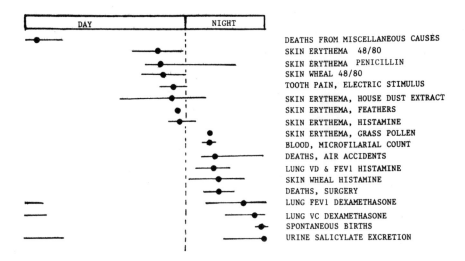

DAY   NIGHT

DEATHS FROM MISCELLANEOUS CAUSES
SKIN ERYTHEMA 48/80
SKIN ERYTHEMA PENICILLIN
SKIN WHEAL 48/80
TOOTH PAIN, ELECTRIC STIMULUS
SKIN ERYTHEMA, HOUSE DUST EXTRACT
SKIN ERYTHEMA, FEATHERS
SKIN ERYTHEMA, HISTAMINE
SKIN ERYTHEMA, GRASS POLLEN
BLOOD, MICROFILARIAL COUNT
DEATHS, AIR ACCIDENTS
LUNG VD & FEV1 HISTAMINE
SKIN WHEAL HISTAMINE
DEATHS, SURGERY
LUNG FEV1 DEXAMETHASONE
LUNG VC DEXAMETHASONE
SPONTANEOUS BIRTHS
URINE SALICYLATE EXCRETION

Figure 10.1   (continued)

have a daily cycle [167, 169, 194] (Fig. 10.1). Peak values occur in urine excretion in the early morning, mental state and physical vigor in the mid-day, weight in the late day-time, deaths from surgery and air accidents in the middle of the night, and births in the late night. We are most sensitive to pain around 5 p.m. and we metabolize alcohol best between 2 p.m. and midnight; alcohol consumption in the evening thus should be less intoxicating than at 6 a.m. [250]

We use *alarm clocks*. In considering entrainment and Zeitgebers in human circadian rhythm entrainment, we have to consider the same cues established for plants and animals—light, temperature, and social cues. However, humans are unique among organisms on this planet in their creation and use of artificial devices to measure time (Fig. 10.2). By using watches and clocks, we may closely time events in our lives. Presumably, the ability to synchronize with other individuals by use of precise timing devices has been advantageous in our evolution. The alarm clock must be considered a potential Zeitgeber for synchronizing human circadian rhythms.

We have analyzed self-reported times of retiring and arising in humans [61]. We studied records kept by 11 graduate students for a month in the autumn and found that the students exhibited a weekend delay of about 2h and a Monday-Wednesday-Friday effect (Figs. 10.3, 10.4). The pattern correlated with the academic schedule. There were no classes on weekends so the students could sleep late by choice. Classes were given in Monday-Wednesday-Friday or Tuesday-Thursday schedules. A longer record was kept by one of us. The record included information as to which days alarms were set. When the alarm days are removed from the record, we can see the pattern exhibited by the individ-

Figure 10.2 Humans are unique in making use of alarm clocks to provide Zeitgebers.

ual—it was erratic but nonetheless synchronized with a 24h environment and not freerunning (Fig. 10.5). It is noteworthy that the pattern of the individual did not exhibit any obvious seasonal cycle (the onsets of flying squirrel activity, in contrast, are closely associated with dusk and vary as photoperiod changes (Fig. 10.6). However, there is evidence that the human melatonin generating system responds to light and to photoperiod. In humans, as in animals, melatonin is high at night and low in the day (measured in urine or blood). Light at night suppresses human melatonin [6, 77, 211], and the human melatonin pattern varies with season of the year [6, 186, 197].

Figure 10.3 Humans control the timing of their sleep-wake cycle. The record is for a graduate student who kept track of the times of daily rising and retiring. The black bar represents the daily awake period between the time the individual arose and the time she went to sleep. The record has been doubled so each line represents 48h of data; the arrow points to the time, 6 a.m., Eastern Daylight Savings Time; the first day was a Wednesday; weekends occurred at (A), (B), (C), (D). The individual synchronized weekday arousals with an alarm clock, but showed a marked weekend delay [61].

Light is an effective Zeitgeber for humans under experimental conditions [109, 181]. Moreover, light pulses shifted the phase of human circadian rhythms [107, 180]. It was possible to entrain humans to artificial 27h and 29h light-dark cycles [382].

Social cues were claimed important for human entrainment in one study, where subjects remained synchronized during four days of complete darkness [18].

It is of practical interest to parents to look at the development of the sleep-wake pattern in human babies (See Figure 2.5) described by Meier-Koll and associates:

> "**1.** During the first 4 weeks, while the circadian rhythm is still quite unpronounced the wake density seems to be organized according to a type of a freely running cycle of 4 h. . . .
> **2.** After the 5th week, since the power amplitude of the circadian rhythm had begun to rise, mean periods of about 6 h occurred during the nocturnal phase, while 4-h periods were still maintained during the day. . . .
> **3.** Finally, after the 11th week, when the circadian rhythm was fully established and completely synchronized with the day and night regimen, the 4-h periods of the ultradian cycle were always accompanied by the diurnal phase . . . the infant sleeps the whole night." [226]

Some abnormalities have appeared in the rhythms of blind human subjects (e.g., freeruns, an unusual phase of the melatonin rhythm [210]), which

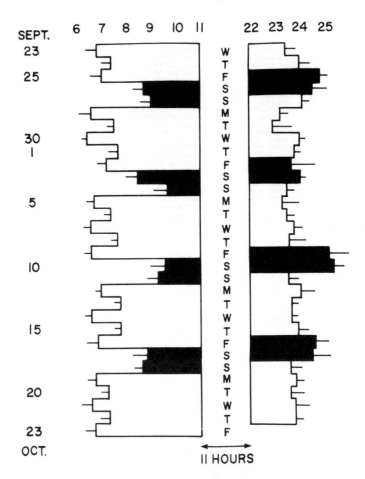

Figure 10.4 Human graduate students exhibit a pattern of rising and retiring times. The data here represent 11 students who reported their own times of rising and retiring. The horizontal axis is time in hours from the prior midnight; dates appear on the left; days of the week appear in the center; Saturday and Sunday appear as black bars on the left; Friday and Saturday appear as black bars on the right. The earliest classes at the university begin at 8:30 a.m. The horizontal lines extending from the bars show one standard error of the mean.

Marked weekend delay in rising on Saturday and Sunday was characteristic of the group and also found in individual records. The group also retired later on Friday and Saturday.

There was weekday effect presumably derived from the students' course schedules. The students rose earlier on Monday, Wednesday, and Friday and rose later on Tuesday and Thursday.

For the group, the average student day lasted 16h and the average duration of sleep was 8h [61].

Figure 10.5 This is a record of a human sleep-wake pattern plotted as in Fig. 10.3, but over a much longer time span. At (A), there was a trip involving a 6h time advance and a 6h time delay on the return a week later. At (B), the individual experienced a change from Daylight Savings Time to Eastern Standard Time (a 1h delay). At (C), the individual traveled west across three time zones requiring a 3h delay for 12 days followed by an advance of 3h on the return to adapt to home time. The record was double plotted; then on the left record, the days that an alarm was used were removed from the record. (Binkley, 1988.)

appear to confirm a role for light as a Zeitgeber for humans. However, other blind subjects were synchronized normally compared to sighted individuals, which is evidence that other cues can synchronize human circadian rhythms.

Longer-than-circadian rhythms exist in humans. One example is the menstrual cycle (Figs. 7.12, 7.13, 7.14). Annual fluctuations occur in such measures as births, especially in the higher latitudes. Ultradian oscillations have been observed in infants and in the occurrence of paradoxical sleep (Chapter 7) [226].

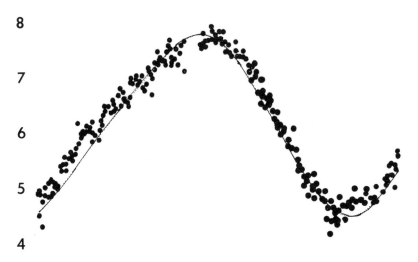

Figure 10.6 Flying squirrel wheel-running onsets (filled circles) synchronize with the time of dusk (line) in natural daylight. The horizontal axis is a year beginning in January. The vertical axis is time of afternoon (CST). (After DeCoursey [120].)

FREERUNS IN CAVES

Freerunning rhythms have been studied in humans by depriving them of time cues in caves [306, 307], in bunkers, or in specially designed laboratories. In looking at these studies, the nature of the time deprivation must be considered—usually the subjects were isolated and kept at constant temperature, sometimes the individuals were permitted to choose the lighting but had no watches, rarely were they kept in DD [13, 306].

Aschoff reports that the freerunning period for 137 human rectal temperatures was 25.00h +/− SD 0.56h [13]. Figure 10.7 shows human freerunning rhythms of activity and temperature. The authors claimed that the activity rhythm *dissociated* from the temperature rhythm so that the activity freerun period was 33.4h and the human freerun period was 25.1h [21]. They called this apparent phenomenon *spontaneous internal desynchronization*. Along with this, in freeruns the internal phase relationships are different than in entrained individuals [21]. In comparing human rhythms with those of most other animals, we run into the problem that the favored experimental subjects have been nocturnal rodents, whereas humans, like birds, are diurnal. Primates, such as the squirrel monkey, also freerun (Fig. 10.8). The period of human freeruns can be slightly altered by light intensity, opportunity to select lighting or temperature, exercise, or a weak electric AC field of 10 Hz [21].

It would seem logical that the *normal* situation for man as well as most other organisms would be to be *entrained* to the prevailing environmental

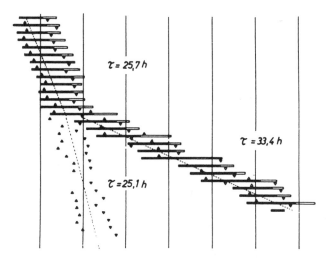

$\tau = 25,7\,h$

$\tau = 33,4\,h$

$\tau = 25,1\,h$

Figure 10.7 Humans exhibit free-running rhythms just as do animals. Here, the black bars mark activity in a human freerunning with self-selected lighting. The vertical lines mark 24h intervals. The small triangular marks show the body temperature cycle. In the study, the authors claim dissociation of the body temperature cycle from the activity cycle. (After Aschoff and Wever [21].)

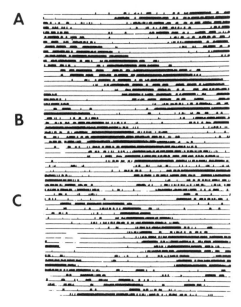

A

B

C

Figure 10.8 Primates have circadian rhythms. The record is wheel-running activity of a squirrel monkey in constant light (LL, 400 lux). The temperature was 20°C in A and C and 30°C during B. (After Tokura and Aschoff [341].)

light-dark cycle, maintaining synchrony with its fellows and the rhythms of the natural world. However, there are reports of individuals who freerun or show relative coordination despite the presence of 24h Zeitgebers (light, social, temperature, mechanical clocks, etc.) in their environment [386]. Nonetheless, the reader should not conclude from the existence of human freeruns that it is necessarily desirable from a health standpoint to live in a freerunning situation. The probable normal situation, the result of evolutionary natural selection, is to be synchronized with the environment and the population.

Individuals who freerun in a 24h world were brought to notice usually because of difficulties they experienced—insomnia (inability to sleep at night) and hypersomnia (day-time sleepiness).

CIRCADIAN SUSCEPTIBILITY

As can be seen from Fig. 10.1, there are also *rhythms of susceptibility*—e.g., to allergens. There is a more dangerous time, early night, to experience surgery. Thus, just as there are optima in chemical and temperature conditions for living things, there are also temporal optima resulting from internal rhythms. Halberg stated this concept:

> "Physiologic regulation has been viewed as the whole of the body's provisions for supplying and transporting the right amount of the right material to the right place. Evaluation of (internal timing) and (external timing of rhythms) adds a new temporal dimension to such regulation, i.e., the right time." [167]

An area where the *right time* concept has broad application is in pharmacology [366]. For example, cyclosporin is a drug from soil fungus used for prolonging the function of organ transplants because of its suppression of the immune system. In rats, this drug prolonged survival for an average 45 days when it was given during the light-time but only 28 days when it was given during the dark-time [219]. It is not unreasonable with current information of time-of-day variation in susceptibility to require that this information be obtained as part of the testing procedure for a drug and published in the *Physician's Desk Reference*. There is a problem with this apparently worthwhile idea, and that is that nocturnal mice or rats might be unsuitable testing material for determination of human daily susceptibility to drugs. In practice, most drugs are administered to humans during the day-time when they are awake, thus potentially imposing a 24h rhythm of drug concentration upon the recipient.

Some circadian scientists have envisioned a more controlled approach to administering drugs in a cyclic manner. Lynch, Rivest, and Wurtman devised an ingenious system for cyclic administration of melatonin to rats that they called *programmed microinfusion*.

> "To infuse melatonin into a rat according to a predetermined temporal program, an Alzet Osmotic Minipump (brand name, from Alza Corp., Palo Alto, Calif.) filled with physiologic saline discharged the contents of a subcutaneously implanted capillary tube that had been precharged with a linearly arrayed infusate program. The program consisted of alternating segments of a melatonin solution ... and ... melatonin-free light mineral oil. The minipump, designed to discharge at the rate of 1 microliter/h when implanted subcutaneously, was attached to the program-containing capillary tubing. . . . Thus, as the contents of

the tubing were gradually displaced by saline solution from the pump, the various components of the programmed infusate were discharged sequentially from the opposite end of the tubing." [217]

The investigators showed that the system worked by observing rhythms in urinary melatonin (Figs. 10.9, 10.10). Cavallini and co-workers have used a different device, an implantable programmable Medtronic® pump to supply a sinusoidal schedule of cyclosporine to beagles [94].

In addition to the rhythms of susceptibility, there is also the problem that some drugs affect the underlying circadian processes themselves. For example, Winget and associates conducted a study with some drugs used as antiemetics (to prevent vomiting) in space flight:

> "The administration of Dexedrine increased the daily mean HR (heart rate) and the amplitude of the HR rhythm. On the other hand, Scopolamine alone reduced the amplitude of the heart rate (HR) circadian rhythm and decreased the daily mean while raising RT (rectal temperature) by 0.5°C above the predrug control mean. Combining the two drugs resulted in an additive effect where Scopolamine reduced the increase in HR produced by Dexedrine alone. In addition, Scopolamine resulted in a major phase shift in one of the four subjects. The data indicate that the administration of drugs in therapeutic doses to healthy humans in a well-regulated environment can affect various characteristics of circadian rhythms and that this may have both useful and hazardous implications in therapeutics." [360]

Clearly, we can improve our use of drugs by paying attention to their rhythms of effectiveness and the consequences they have for our circadian systems.

DEPRESSION AND CIRCADIAN RHYTHMS

Figure 10.1 lists brain and performance peaks occurring around the mid-day for humans. Many of the rhythms studied, for example, locomotor activity, are considered to be *behavioral* rhythms. In 1965, Curt Paul Richter [282] discussed periodic physical and mental illnesses. His anecdotes describe cyclic psychotic attacks, a schizophrenic whose personalities occupied alternating days, periodic dipsomania (drinking sprees), and a 20-day attack cycle in a catatonic-schizophrenic. Luce's [215] discussions of rhythms and psychiatry include the potential relationship of rhythms to depression, and, in animals, rhythms in brain amines (e.g., in norepinephrine in the hypothalamus, pineal, lateral thalamus, mesencephalon, pons, and cervical spinal cord; e.g., of serotonin in the red nucleus, hypothalamus, and telencephalon), memory, and conditioned responses. Luce, as well as Sack and co-workers [300], include the menstrual syndrome (now called PMS for premenstrual syndrome) among periodic illnesses. In particular, periodic oscillations in mania and depression

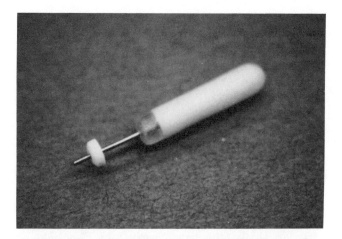

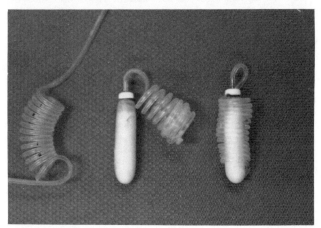

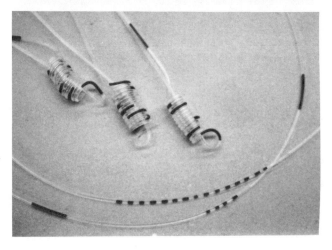

Figure 10.9 A mini-infusion pump (Alzet®) can be used to deliver drug doses on a circadian schedule. The minipump contains a balloon that is filled with liquid. Placed in an animal's body, the white portion of the pump absorbs fluid from the organism; the fluid presses on the balloon delivering the liquid at a constant rate. To make the pump administer a circadian schedule, a long plastic tube is filled with a *program* of the desired drug (e.g., melatonin) dissolved in saline water alternating with an oil (e.g., melatonin-free mineral oil) that does not mix with water. The tube is attached to a minipump and coiled around it. The timing of the program can be varied by changing the amounts of tubing occupied by the drug solution and by the oil. When the minipump is implanted and begins to work, it forces the solutions out the tubing in order. The technology, called *programmed microinfusion*, was developed by H. Lynch, Rivest, and Wurtman [217].

The photographs show: (A) the Alzet minipump with its filling tube and cap; (B) the tubing coil by itself on the left, attached to the minipump in the center, and in position around the pump on the right; (C) uncoiled and coiled tubing containing the solution programs.

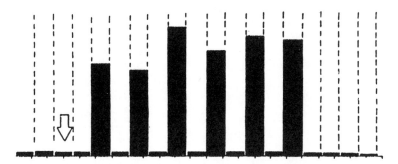

Figure 10.10 Illustration of rhythmic melatonin produced in a rat using the technique of programmed microinfusion (Fig. 10.9). The dashed lines divide 12h intervals, the bars represent melatonin. The pump was implanted at the arrow. (After Lynch, Rivest, and Wurtman [217].)

were noticed and circadian rhythm disturbances were associated with manic-depressive illness [207, 373].

More recently, the relationships of light, depression, and rhythms have attracted notice. Psychiatrists now have a name for a disorder, SAD (Seasonal Affective Disorder), which is described by Sherer and co-workers as:

> "a condition characterized by recurrent depressions in the fall and winter interspersed with non-depressed periods in the spring and summer. In their winter depressions, patients with SAD complain of fatigue, sadness, oversleeping, overeating and carbohydrate craving. They withdraw from other people and relationships invariably suffer. Frequently, these patients complain of cognitive difficulties and impaired productivity." [305]

Sack and others [300] list symptoms characteristic of SAD (Table 10.1). In their report, 86 percent of SAD patients were female, 69 percent of the patients had a family history of affective disorder, and 93 percent of their sample had a lifetime diagnosis of bipolar disorder. The existence of SAD appears to be verified by a report of a similar syndrome in the southern hemisphere (Australia) [78]. While a winter-summer difference could be due to environmental factors other than photoperiod (such as seasonal weather changes), investigators have focused on light and have suggested that SAD is related to the short photoperiod of winter.

Perhaps the most amazing outcome of the application of circadian rhythm principles (e.g., an understanding of entrainment and photoperiodism) to SAD was that light has been used successfully to treat the depression of SAD patients [187, 206, 207, 209, 206, 298, 309]. Basically, the idea was to increase the short winter photoperiod by the addition of artificial bright light exposure (e.g., 2500 lux). A number of investigators have had positive results (e.g., remissions of depression in 87 percent of patients) within a few days;

TABLE 10.1 SAD Symptoms

Sadness	Later waking
Anxiety	Increased sleep time
Irritability	Interrupted, not refreshing sleep
Decreased physical activity	Daytime drowsiness
Increased appetite	Decreased libido
Carbohydrate craving	Difficulties around menses
Increased weight	Work difficulties
Earlier sleep onset	Interpersonal difficulties

[1]Symptoms observed in more than 66 percent of patients [300]

relapses followed light treatment cessation. The methods of *phototherapy* treatment varied among investigations (e.g., pulses in the morning, pulses in the evening, a skeleton of twice daily 3h pulses at 7:30 a.m. and 8 p.m. [372], 5–6h of bright light/day [300]), but the results taken together have been that added light consistently is capable of ameliorating depression.

There is evidence that manic-depressives have disturbed circadian rhythms [207]. Wehr and co-workers [373] review circadian rhythm abnormalities in victims of manic-depressive illness and identify three types: (1) reduced amplitude of some rhythms; (2) advanced phase of some rhythms; (3) the occurrence of 48h sleep-wake cycles in association with the transition from the depressed state to the manic state. Moreover, clinical manifestations of manic-depressive illness include early morning awakening and diurnal variation in mood [373]. Wehr and his colleagues noted that four of seven manic-depressives advanced their wake-up time when they came out of the depressive phase of their illness [374]. The investigators were able to obtain an antidepressant effect in one individual by advancing her sleep-wake schedule by 6h.

In seeking markers and a biological basis for the effects observed, investigators have used a *melatonin hypothesis* based on the light regulation of melatonin biosynthesis that was discovered in animals and confirmed in humans. A piece of evidence that supports the idea that melatonin may be involved is that manic-depressive patients were supersensitive to light (e.g., 500 lux at 2–4 a.m.) [209]. The melatonin rhythm in depressed patients, like some other rhythms (e.g., cortisol), was abnormal in some studies (e.g., lower amplitude [102], reviewed by Vaughan [357]). Melatonin administration worsened depressed patients, lowered oral temperature, reduced sleep time, and affected reaction time [305, 357]. However, the fact that light in the day time reduced depression, and the fact that atenolol (which lowers melatonin) only reduced depression in three of 19 SAD patients, have been used to argue against the melatonin hypothesis [297].

While the nature of the light source used to treat depression may not matter, as long as it is bright, most investigators have used full spectrum lighting typified by the Sunbox®, which provides 2500 lux at three feet using Vita-lite® tubes.

I speculate that the potential role of light and circadian rhythms in affecting mood may not be limited to SAD patients, but may be a factor in the lives of many of us who live at temporal latitudes. Manipulation of light quality and timing in the environment—even by simple measures such as light-colored rooms, bright lighting fixtures, winter vacations in sunny locations, etc.—may be important at temporal latitudes.

Sack and co-workers [300] and other authors include PMS (Premenstrual Syndrome) in the subject of biological rhythms in psychiatry. Symptoms of PMS include (compare them to the SAD symptoms in Table 10.1) [161,300]:

irritability	energy disturbances
depression	weight gain
anxiety	breast swelling
difficulty concentrating	headache
negative self-perception	bloating
sleep disturbances	edema
(hypersomnia)	emotional lability
increased appetite	constipation

In the article, the authors link PMS to other depressive disorders and SAD. Theories about the cause of PMS have to do mainly with an alphabet soup of hormone changes—progesterone deficiency relative to estrogen, prolactin increase, mineralocorticoid increase, high prostaglandins, decline in endogenous opiates, biogenic amine metabolism (serotonin, dopamine, norepinephrine, acetylcholine), and menstrual variations in glucocorticoids or melatonin.

JET LAG, ADVICE FOR THE TRAVELER

When we travel east or west across time zones, we are required to resynchronize to the time at the destination. East-west travel thus requires a phase shift of the circadian biological clock (Fig. 10.11). The term *jet lag* has been defined by Palmer as "a melange of symptoms, dominated by a disrupted sleep pattern, occurring when one's physiological rhythms are out of phase with the ambient light-dark cycle (and each other) after transmeridional flights" [250]. However, it has not been fully appreciated that traveling north and south in the same time zone does not require a phase shift, but (if there is a change in the photoperiod so that the times of dawn and dusk are different) may require resynchronization and produce the symptoms associated with jet lag (Fig. 3.10). In considering this subject, it is impossible to separate the symptoms of circadian rhythm disruption from other possible physiological consequences of journeys for human travelers as it is for controlled subjects subjected to phase shifts of their Zeitgebers in the laboratory.

Ehret and Scanlon [141] describe symptoms of jet lag—exhaustion, loss of ability to concentrate, performance decrements, constipation or diarrhea, insomnia, loss of appetite, headache, impaired night vision, limited peripheral vision. Recovery rate differs among individuals and differs for the various parameters; however, Ehret and Scanlon [141] suggest that it takes two days to two weeks to make a 5–8h time change. As Luce notes: "Since rapid travel has become a way of life for a sizeable population, it would seem urgent to learn how to accelerate resynchronization" [215].

Aschoff and his colleagues reviewed the literature on transmeridional flight and concluded that the bulk of experiments supported a faster re-entrainment after east-to-west travel than after west-to-east travel (the *asymmetry-effect*) [19]. However, some investigators did find eastward re-synchronization faster, as did the individual whose record is shown in Fig. 10.11. The results for travelers contrast with data for shifting in isolation chambers where the advancing rate was 70 min per day and the delaying rate was 54 min per day [19]. The authors of the review speculate that the difference in the isolation and travel results may be that "in the isolation experiments so far the subjects had not been informed about the purpose of the study; they therefore had been surprised by the change in lighting regimen" [219]. There were other differences as well in social contact, the difference between artificial (weak?) and natural (strong?) Zeitgebers, stress of traveling, and the sleep deficit that is incurred by travelers. However, the figures lead to a crude rule of thumb of about a day of readjustment required for each 1h change in time zone. The circadian researchers have also offered an explanation for the malaise of jet lag in their observation that different rhythms of an individual's body phase shift at different rates so that during readjustment, the person's rhythms do not have their normal phase relationships (internal desynchronization). There are many laboratory studies of animals regarding the rates of phase shifting. Sparrows were subjected to repeated advancing or delaying 8h phase shifts and their rates of re-entrainment were measured. The fastest re-entrainment occurred if the sparrows were advanced and if constant light (which produces increased alpha in sparrows) intervened between the pre-shift and post-shift cycle (Figs. 10.12, 10.13) [57, 60, 63].

I think that it should be possible to discover the optimum conditions to produce the fastest possible resetting (and minimum jet lag malaise) for any given trip for an individual. In other words, it should be possible to make a *prescription for jet lag*. I visualize this prescription to consist mainly of recommendations for lighting and sleep-wake schedules prior to and during the days of travel. Other researchers have envisioned schedules involving diet [141], and drugs (caffeine [141], benzodiazepines [345], and melatonin [7]). Let us discuss, then, some possible strategies for minimizing jet lag.

First, a simple strategy is to *remain on home time*. That is to wake and

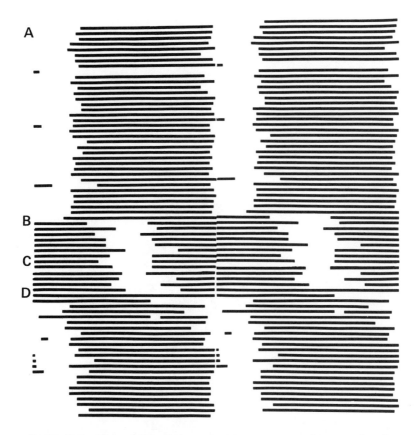

Figure 10.11 Phase shifting is important to two aspects of human circadian physiology. Phase shifts are required in east-west travel and in adapting to shiftwork schedules. The record shows a traveler on Eastern Daylight Savings Time before his journey (A), while he was halfway around the world in Japan (B) and Hong Kong (C), and after his return to his home time zone (D). The record is double plotted on Eastern Daylight Savings Time (horizontal axis = 48h) and the lines representing successive days are arranged vertically in chronological order. (Binkley, 1988.)

retire at the usual times with respect to time in the home time zone. This strategy is often feasible for short trips involving only a few time zones. Using this strategy may take some preplanning in scheduling at the destination so that meetings would take place at a time that is comfortable in the home time zone. For example, traveling 3h west (e.g., from the east coast to California), the earliest meeting time might be scheduled for 11 a.m. California time to coincide with 8 a.m. Eastern time.

Second, a strategy is to *preadapt to the new time zone* before the trip to the new time zone or in stages. This strategy is particularly desirable if important meetings are to take place at the destination. Preceding the trip, one simply

18 6

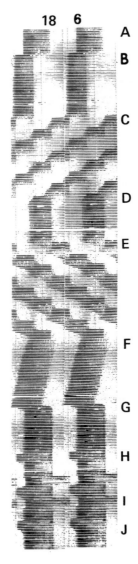

A

B

C

Figure 10.12 Sparrows readily phase shift and re-entrain their circadian rhythms. The figure is a record of a house sparrow subjected to a sequence of light-dark schedules; the record is doubled so that one line is 48h of data. In (A), the sparrow was entrained to LD12 : 12 (lights-on 6 a.m.) At (B), the sparrow was placed in LD8 : 16 (lights-on, 2 a.m.). The sparrow was subjected to eight phase shifts (B, once a week, 8h advanced) in which constant dark was imposed between each schedule for two days. The sparrow was left in LD8 : 16 (D) and then subjected to nine phase shifts (E, once a week, 8h delayed). At (F), the sparrow was left in DD where it freeran until it was placed in LD12 : 12 (G). IN (H)–(I), the sparrows had two light bulbs which were programmed with separate schedules. One light was LD12 : 12 (lights-on 6 a.m.) and a second light had its schedule rotated [as in (C) and (E) above]—delayed in (H) and advanced in (I). LD12 : 12 alone was restored at the end (J). (Binkley and Mosher [57].)

begins to rise and retire (using destination time zone times) at the times one desires to wake up and go to sleep at the destination. The farther the trip, the longer the preadaptation period would be required. This approach may be impractical for most trips, but sometimes partial preadaptation could be achieved (preadapting for a 3h change in anticipation of a 6h time change on a trip).

Third, another strategy is to *make use of the weekend lag effect.* You can make use of this effect if you have weekend lag, and you know how much later

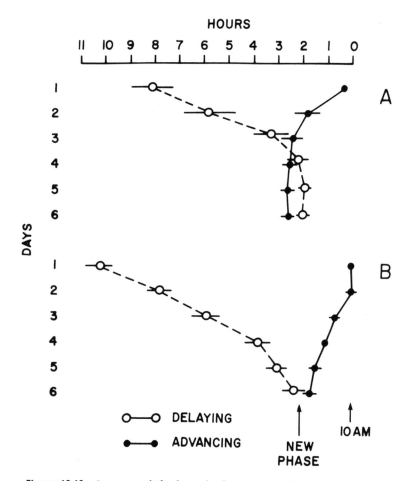

Figure 10.13 Sparrows shift phase the fastest when they are advanced and constant light intervenes between two schedules. The graphs show the rates with which sparrows re-entrain to a schedule of LD8 : 16 shifted by 8h. (A) shows re-entrainment of sparrows' activity onsets if constant light intervened for two days between schedules; (B) shows the re-entrainment when constant dark intervened; open circles are adaptations to delayed schedules; filled circles are adaptations to advanced schedules. The measurements are for five sparrows/point as the sparrows re-entrain to a new schedule that had lights-on 10 a.m. Sparrows normally entrain to LD8 : 16 with anticipatory activity (Fig. 3.3) represented by New Phase [63].

you normally sleep on Saturday than on weekdays. For example, here are recommendations for a person with a rhythm such as that shown for the individual in Fig. 10.3. The example is for a trip from the east coast to the west coast of the United States. Get up at your normal weekend time on Sunday, travel west; go to bed on west coast time on Sunday; on Monday wake up using west coast time. If you have a typical 2h weekend lag, you would have only a 1h adjustment to adapt to California time. This phenomenon may in fact account for the asynchrony west-is-faster-than-east effect found in many studies of transmeridional flight. Return to the east coast on Friday; go to sleep on east coast time on Friday; you will awaken early Saturday and Sunday (when you normally get up late) but close to comfortably on time by Monday (when you normally rise early).

Fourth, *stay awake from departure until bedtime at the time of the destination*. This is the strategy that was effective for sparrows. This strategy has the advantage that it is simple to remember. Stay awake and keep the lights on during the trip until the desired retiring time in the destination time zone. Go to bed and, even if you don't sleep well, wake up and begin your day's activities at the desired time in the destination zone.

Fifth, *use Scanlon and Ehret's Jet Lag Program* [141]. This program, which they spell out explicitly for different travel plans, uses a combination approach: (1) It uses the first strategy above (using lighting and physical activity anticipating the day-time of the new time zone). (2) It includes food (high protein breakfasts and lunches for long-lasting energy, carbohydrate dinners for encouraging sleepiness). (3) It uses drugs (methylxanthines, obtained as theophylline in tea; or caffeine and theobromine in coffee or in chocolate, or in nonprescription over-the-counter drugs). The program is not simple and it requires a feasting-famine schedule before the trip; however, its proponents claim it can reduce jet lag to three days even for a 12h time change.

Sixth, it may be possible to *use a pharmacological strategy*. Investigators and the pharmaceutical industry are interested in possible drugs that could be used to treat jet lag. The work has followed two approaches. The first is the attempt to alleviate jet lag with melatonin [7], whose exogenous administration can cause fatigue and sleep [387]. The second has been the administration of benzodiazepine (triazolam, a drug for insomnia) to phase-shift animals [345, 346]. As Winfree put it:

> "The potential practical interest . . . lies in the hope that . . . a fast-acting, quickly eliminated chemical without conspicuous side-effects may also abruptly reset the circadian clock in humans. If so then we may soon find ourselves carrying pills and a schedule whenever we cross time zones . . . The schedule will prescribe when and what dose of pills to take to achieve an immediate advance or delay appropriate to the foreseen change of longitude." [385]

Faced with a plethora of solutions, however, the reader may throw up his hands and choose to agree with one of Winfree's remarks: "My own favoured procedure involves exposure to bright sunlight at a carefully planned hour" [385]. As we have seen, this approach might not be altogether ineffective. This is the approach, after all, that is used by migrating animals.

SHIFT WORK, A CIRCADIAN PROBLEM

According to Czeisler, almost 27 percent of the United States work force is employed in *shift work*, that is, work that is scheduled during all or part of the night [108]. Some of the individuals on shift work schedules encounter similar problems to those of travelers. Whereas the travelers' adjustments are reinforced by the natural schedule at their destinations, the shift workers work at unusual (non-daytime) phases in an environment that provides conflicting rather than reinforcing social and lighting time cues. In addition, the shift workers, unlike occasional travelers, experience repeated disruptions. Moore-Ede and Richardson reviewed the medical implications of shift work and note that they include:

> " . . . sleep-wake disorders, gastrointestinal pathology, and an increased risk of cardiovascular disease. There is significant interindividual variation in the ability to adapt and also a deterioration with age. Evidence is accumulating that poor adapters present with a Shift Maladaptation Syndrome with characteristic pathological manifestations . . . relate(d) to a loss of internal coherence in the multioscillator circadian timing system." [237]

Another description of the consequences of shift work was offered by Reinberg and his colleagues:

> "Clinical intolerance to shift work was defined by the existence and intensity of a set of medical complaints: sleep alterations, consisting of subjective self-ratings of poor sleep quality, difficulty in falling asleep, frequent awakenings, insomnia, etc; Persisting fatigue which did not disappear after sleep, weekends or vacations, thus differing from physiological fatigue due to physical and/or mental effort; changes in behavior consisting of unusual irritability, tantrums, malaise, and inadequate performance, etc; digestive troubles ranging from dyspepsia to epigastric pain and peptic ulcer. The final and almost pathognomonic indication of intolerance to shift work was the regular use of sleeping pills of any kind (barbiturates, benzodiazepines, tranquilizers, etc.)." [277]

Just as for travelers, there are a number of scheduling strategies that can be adopted which may reduce the consequences of shift work that involve circadian rhythm principles.

First, and simplest, is the fixed schedule strategy in which a worker is hired to work a specific shift. It is possible, for example, to schedule two 8h shifts during the normal day time (e.g., 7 a.m.–3 p.m., 3 p.m.–11 p.m.), permitting sleep during the night, and one shift in the normal sleeping time (e.g., 11 p.m.–7 a.m.). This system depends on employee preferences, and pay incentives can be used to fill the shifts. It would seem from the weekend-lag effect that it might be easiest for most workers on the 11 p.m.–7 a.m. night shift to do their sleeping in a darkened room from 7 a.m.–3 p.m. An individual's entrainment to the cycle could be enhanced by the use of bright light in the workplace, by adherence to a regular schedule, by use of methyl xanthines, and by meal timing and content. From a circadian point of view, the fixed schedule method seems potentially healthiest because it should be possible to achieve stable entrainment and internal synchronization albeit with an unusual phase angle. However, employers have developed systems in which responsibility for shift work is shared among all employees by using *rotating* shift work schedules.

Second, there are advancing rotated shift work scheduling strategies. Czeisler and co-workers describe one such schedule, and Conroy and Mills discuss the implications:

> "Weekly shifts were rotated with each crew working a given 8-hour shift for 7 days before rotating to the preceding 8-hour shift. Hence the scheduled work time rotated in a phase advancing direction from night (midnight to 8 a.m.) to swing (4 p.m. to midnight) to day (8 a.m. to 4 p.m.)." [108]

> "If one tries to recommend the best arrangement for shift working, the physiological and social considerations conflict. Teleky . . . on the basis of the time taken for the temperature curve to adapt to night work, recommended for factory workers a rotation at longer than weekly intervals and preferably at monthly intervals, and for other workers, such as nurses, rotation at even longer intervals. The usual custom at present in many industries, to work about a week on one shift, have a rest break, and change to another shift, is at complete variance with this recommendation. It is perhaps the most unsatisfactory arrangement that could be devised, for just as the subject's rhythm has been thoroughly disturbed and he is beginning to entrain to a new routine, he is switched to a different schedule . . . . Either extreme is better, though for different reasons. Continuous working on night or any other shift permits the maximum of adaptation of all physiological rhythms, particularly in an isolated community such as the Spitsbergen coalminers where the social life is geared to the working routine. For most people living in a community who follow regular nychthemeral habits this is, however, unacceptable on social grounds. [105]

In most industries, the rotated schedule is interrupted by days off, weekends off, and vacations, during which the individual is free to select his own schedule. It is not clear what the best schedule is from the standpoints of worker satisfaction and health.

We subjected sparrows to rotated schedules using light-dark cycles (Fig. 10.12) to study the effect on their circadian rhythms [57, 63]. Sparrows readily re-entrained to repeated phase shifts of five days of LD8:16 whether it was advanced or delayed (Fig. 10.12). We also did the experiment where the sparrows experienced DD for two days between the shifts or two days of LL between the shifts. When we compared the rates of re-entrainment in the sparrows, we found that they re-entrained fastest if they were subjected to advancing rotations with intervening LL (Fig. 10.13). We can explain this with the sparrows' PRCs for light and dark pulses because they have larger advancing than delaying portions.

Third, there are delaying shift work scheduling strategies. Czeisler and co-workers [108] studied rotating shift workers and found that worker work schedule satisfaction, subjective health estimates, personnel turnover, and worker productivity were improved by use of delaying schedules and less frequent rotations (every 21 days) as compared to weekly advancing rotations. They attribute this to the fact that the human freerun is about 25h.

Fourth, the use of artificial lighting can be incorporated into the first three strategies. Wever studied the relationship of light intensity and phase-shifting in humans [381] and concluded that a light intensity of 3000 lux (the normal range of artificial illumination is less than 1000 lux) is required to be effective in synchronizing human rhythms. As regards shift work, he suggested that: " . . . bright light during the working hours, which has been shown to affect particularly the physiological functions, may lighten the burden" [381]. I cite another study to emphasize the importance of light. Broadway and Arendt [82] measured the recovery of melatonin and sleep rhythms after night shift work in the antarctic. During the winter, when there is a prolonged lack of sunlight, night work phase-shifted the melatonin peak from its normal night position to a day position. But at the vernal equinox, when there is a pro-nounced light-dark cycle, the melatonin rhythm occurred during the normal night-time. Either night work was insufficient to shift the melatonin rhythm in spring or it rapidly re-entrained.

Taken together, what we know about shift work leads to the following recommendations if rotated schedules are to be used. First, the frequency of rotation should be as low as possible. Second, bright light should prevail in the workplace. Third, delayed rotations should be used. Fourth, in changing from one shift to another, strategies such as those suggested for travelers should be used (e.g., making use of the weekend delay effect). However, Turek, my colleagues, and I have pointed out that the rotated shift worker does not neces-sarily follow the rotating schedule as do the sparrows. The rotated worker is subjected to two simultaneous schedules: the shift work schedule, and the more constant schedule prevailing in the environment. When the sparrows were subjected to two simultaneous schedules (Fig. 10.12H, I), LD8:16 rotated weekly and LD12:12 fixed, they did not follow the rotations. Instead, they responded as though they were being subjected to a series of photoperiods. As Turek put it:

" . . . delaying or advancing the work time does not imply that the sleep-wake cycle is also shifted in a similar manner. Indeed, after a complete rotation between the day, evening, and night shifts the sleep time will be advanced once, delayed once, and not shifted once, regardless of whether the workers are on a delaying or an advancing work rotation schedule. Thus circadian rhythms are likely to be perturbed in a similar manner whether the work schedule is rotated in a delaying or an advancing direction." [344]

Other investigators concluded that night-time naps compensated for the loss of sleep that occurs in night shift workers [221]. Clearly, more study is needed. For example, it may be possible that productivity may so wane during the course of a night shift that an employer might reap increased productivity by having two 4h night shifts and paying a higher hourly rate.

CIRCADIAN RHYTHMS AND WORK IN SPACE

The reader should by now appreciate the fact that circadian, circannual, and lunar-tidal rhythms have evolved because of the daily and annual fluctuations in the lighting and environment on earth due to its rotation rate, orbiting rate, and its relationship to the sun and moon. So it follows that voyagers into space are presented with two problems. First, a spacecraft orbiting the earth every 100 minutes experiences a 100 min light-dark cycle, not a 24h cycle, and less than 30 percent of time is spent in the earth's shadow (dark phase). Second, *daylength* on other planets in our solar system is not 24h (for example, about 15 days of sunlight alternate with 15 days of night on the moon; Mercury's rotation is 59 days, Venus rotates in 243 days, Mars rotates in 24h 37 min, Saturn and Jupiter rotate in 10h, Uranus rotates in about 24h, and Pluto rotates in a little over 6 days [251]). Third, during a flight such as to the moon, travelers would be subject to a black sky and to constant sunshine. Fourth, the light intensity on other planets is different than on earth, for example, on the moon the earthshine is 75 times the illuminance of the full moon [105]. The traveler seems to have infinite scheduling options. In fact, his circadian physiology limits these options.

The impact of using the prevailing lighting in a spacecraft or on another planet can be predicted from some animal and human studies described in this book. First, from the animal studies we can make two predictions for the outcome if the astronauts subject themselves to the 100 min cycle—activity during the light phase or freerunning through the cycle (Figs. 3.2 and 3.17). Second, on a planet such as the moon, the spacemen would experience prolonged periods of constant light or constant dark, somewhat as occurs in the arctic and antarctic; we know from one study [82] that the effects of night shift work are altered under antarctic conditions. We know also that human circadian rhythms would freerun in constant light and dark, but we would expect social synchronization in a colony of space adventurers. Third, if the space travelers were permitted ad libitum (self-selected) lighting, we would predict that they would choose a circadian pattern (Fig. 2.2) [59, 341]. Fourth, on the basis of T experiments (see Section 3.3), we would predict that humans could readily adapt to cycle lengths within their range of entrainment. Wever [378] discusses data for human entrainment to T cycles and shows an upper range in the vicinity of 27h and a lower range in the vicinity of 22.5h, but points out that individual freerun periods (145 of 152 human subjects) had tau between 24.0 and 26, median 24.9. Fifth, as discussed in this chapter, the intensity and timing of prevailing light might have a role in mood and circadian rhythm control.

One organism that was actually sent into space is the fungus, *Neurospora crassa*. Race tubes (Chapter 1) were packaged in foam and sent aboard a space shuttle. When the cultures were reexamined upon their return to earth, rhythms with period about 22.7 h were observed, so the freerunning rhythm persisted in space. The cultures were affected by the spaceflight (which involved large temperature changes among other disturbances) and were "visibly different" from the control cultures grown on terra firma:

"1. there was a much greater variation in growth rates among the tubes
2. there was an increase in variance of the circadian period
3. the clarity of banding, which probably reflects the amplitude of the circadian rhythm, was considerably reduced" [318b]

What actually happens to space travelers? Coleman has commented upon this:

"Astronauts have repeatedly complained about poor sleep, sleep loss, and the accumulation of fatigue during their missions. The space traveler is confronted by an array of new environmental stimuli potentially disruptive to the biological clock: long working days or missions, absence of recurring 24-hour time cues, shifted sleep and work schedules in comparison to ground control, as well as noise from round-the-clock radio transmissions—not to mention weightlessness and loss of gravity. Brain-wave recordings of Skylab crew members revealed that some astronauts slept less in space, but most actually slept for the same length of time they did on earth." [104]

Conroy and Mills [105] also discuss the experience of astronauts and mention lost sleep, fatigue, and reduced performance as potential consequences of travelling in a space vehicle. Russian cosmonauts sleep when it is night in Russia, and some American crews have slept when it is night in Cape Kennedy, thus maintaining their home time schedules [105].

Rhythms persist during space travel. The heart rate rhythm of command pilot Frank Borman had a 23.7h pattern when the astronaut lived on a 23.5h day on the Gemini VII mission [250]. Luce quotes Soviet cosmonaut Titov on the nature of sleeping in zero gravity:

"Once you have your arms and legs arranged properly space sleep is fine. There is no need to turn over from time to time as a man normally does in his own bed. Because of the condition of weightlessness there is no pressure on the body;

nothing goes numb. It is marvelous; the body is astoundingly light and buoyant."
[216]

It seems safe to assume that disruption of established circadian patterns in human space travelers would at least have some of the deleterious consequences encountered by earth travelers and shift workers. Moreover, it seems that similar solutions to these problems would pertain in space. In making preliminary recommendations for space travelers, I would suggest:

1. Remain on the home 24h time sleep-wake and meal schedule (unless on a planet whose rotation rate is within the range of entrainment, e.g., Mars).
2. Preadapt before lift-off for shift work schedules.
3. Use very bright light during the planned subjective day and provide for a darkened, quiet sleeping area.

MAKING THE MOST OF TIME

It seems to me that there may be much to gain in daily life by a less casual consideration of the role of circadian rhythms in our lives. It is clear from Fig. 10.1 that we can expect optimum activity, that we can make use of peak of mental activity and physical vigor that occurs about 9h after rising (about 4 p.m. for a 7 a.m. riser). Presumably the actual peak for an individual could be determined and used in making schedules. Mid-night air travel and surgery are obviously things to avoid.

For an individual, it should be useful information to know his own activity pattern. Human activity patterns have been measured in laboratories, but there may soon come a day when Human Activity Monitoring Systems (HAMS or "biological watches") are available for as routine use as are wristwatches. Without data, we self-describe ourselves as larks or owls (of the students I studied [61], six described themselves as owls and three described themselves as larks). With activity monitoring, the actual pattern could be determined and an individual could schedule his or her tasks to correspond appropriately with the times of high and low activity. The information, together with data collected during a phase shift (shift work or travel), might make it possible to select individuals best suited for particular shifts or for shift work in general. To the extent that activity is a measure of health, HAMS might be useful in evaluating drug effects and the efficacy of other medical treatments.

One of the first pieces of data I encountered when I entered the field of circadian rhythm research was the activity record of a sparrow kept in constant light (Fig. 5.3). While my personal time was shortened by the need for nightly sleep, here was a tiny animal that could hop 24h per day for an indefinite length of time with no apparent ill effects. And this dramatic ability was evoked by nothing more complicated than keeping the lights on. To me, the possibility of decreasing sleep need and having a longer time per day in which to be conscious, to enjoy life, is one that is intriguing. Through the manipulation of circadian rhythms (with brighter light in the environment? with extended bright light photoperiod? with light pulses? appropriately timed naps? with appropriate drugs?), it seems possible to me that this goal could be achieved.

Coleman [104] suggests for space travelers the possibility of living on a 25h daily sleep-wake schedule. What would be the likely circadian consequences? Investigators using animals [16, 150] have shown that as tau increases, phase angle increases. Thus, on a 25h day, an earlier wake-up time with respect to lights-on is expected. The effect of a non-24h regimen upon the timing of activity onset and the distribution of activity can be quite dramatic. For example, Aschoff describes the effect of a 26h day on nocturnal white mice:

Figure 10.14 The figure represents some of the adjustments humans are required to make in the phase of the timing of their activity (16h of hypothetical activity are represented by the black horizontal bars). The vertical lines are 24h apart. First, we make a 1h advance to change from Daylight Savings (A) to Standard Time (B). Second, we delay our clocks 3h when we travel from the east coast time zone (C) to the west coast time zone (D) of the United States. Third, a worker using 8h delayed rotations makes repeated 8h delays as he changes shifts from (E) to (F) to (G). When we freerun with a 25h period or live on a 25h schedule, the pattern is that in (H).

Figure 10.15 An individual seated before a Sunbox®, which provides bright full-spectrum light. Some studies indicate that such light exposure can reduce psychological depression.

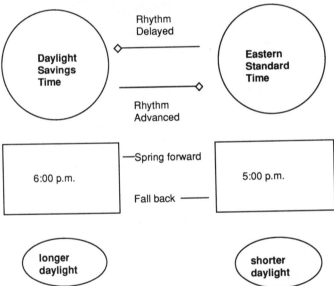

Figure 10.16 The diagram shows the annual time changes humans make in changing between daylight savings time (left) and standard time (right).

"In a 26-hr day, *onset* of activity as well as maximal activity is advanced relative to its position in a 24h day, so that the normally dark-active animals become practically light-active. In the 22-hr day, the biological phases are delayed, the animals now being even more dark-active than in the 24-hr day." [11]

In other words, treatments that change phase angle (photoperiod, T cycles) with respect to lights-on or lights-off have the potential for turning larks into owls or owls into larks, which might be useful in planning shift work. Wever [379] reported a reduction of the percentage of time spent sleeping and absolute amount of time spent sleeping when tau was longer in humans (e.g., 7.7h for 25.47h tau males, 8.34h for 24.57h tau males); human females had a larger fraction of sleep and a shorter period than males. Moreover, Wever [378] suggested individual variation in the range of entrainment for humans. Clearly, the circadian system is subject to manipulation. On the other hand, as a species we have adapted for our entire existence to a 24h day, so that use of a non-24h regimen should be undertaken with caution.

In fact, we do make some routine use of manipulation of scheduling in our lives. Each spring as the days lengthen, we shift ourselves to Daylight Savings Time (we set our watches 1h forward, we thus delay our rhythms by 1h). Then, each autumn, we change to Standard Time (we set our watches back 1h, we thus advance our rhythms 1h). However, the effect of this change is barely discernible in the record of Fig. 10.5. In my opinion, it is only dawn of the morning of our use of our circadian physiology to improve the quality of our lives. As reiterated in Fig. 10.14, there are a number of problems related to circadian rhythms that we routinely face in our daily lives. In the work with depression (SAD), we are initiating efforts to manipulate our circadian rhythms for our own benefit (Fig. 10.15). Many of us routinely alter our annual schedule to match the progression of the seasons when we change between standard and daylight time (Fig. 10.16).

EPILOGUE

The reader may be a bit unsatisfied. At the outset of this book, I posed the question as to whether time should be considered a *sixth sense*, which I leave for the reader to consider. The ability to keep track of various kinds of time, certainly to oscillate physiologically, is a property of organisms. An interesting aside comes from investigations of the abilities of humans to estimate time intervals. Remarkably, this ability is not temperature compensated; it is dependent upon body temperature. When the body temperature is high or the thyroid gland overactive, folks estimate a shorter minute and perceive an actual minute as a longer period of time. Since body temperature has a circadian rhythm, it may not be surprising that the ability to estimate time intervals has circadian variation. Subjects who have been permitted to freerun, in caves underground without mechanical clocks, have underestimated the passage of time.

I have also not really answered the question of why organisms possess the ability to freerun. However, if that ability is accompanied by the capacity for prediction of changes in the environment, it seems logical that it would have adaptive value. It also seems to me that the ability to freerun, and a rhythm in resetting sensitivity, may be the mechanism evolved that permits resetting and maintains entrainment. As pointed out by many investigators, for example by Alleva [1], one reset by one signal per day is theoretically more than sufficient to maintain synchronization with a 24h day, yet many investigators, including myself, have come up with models in which more than one resetting signal (e.g., both dawn and dusk, more than one Zeitgeber) are used.

I also haven't really shown you a clock, though we have some structures, some cells and brain regions, that have the ability to generate cycles. Also, we have some potential clockparts in nucleic acids, membranes, and enzymes. If the lack of a visible clock is disappointing today, consider that most of the people in the *field of biological clocks* are alive today and vigorously competing to find the clock mechanism(s), so that in your lifetime you have a good chance of participating in a scientific *eureka!*, the one in which the circadian clock is finally elucidated. I believe there is a basic mechanism, contained within single cells, common to many organisms, that is responsible for circadian pacemaking. Moreover, I think that when it is revealed it will be simple and elegant.

In 1970, when I was just embarking on my own quest to study the circadian clock, one of the astronauts, Donald Holmquest, wrote me:

> "If you intend to do serious research in this area as part of graduate work, then my only advice is to be very careful. The field of circadian rhythms is one fraught with difficulty and laden with mythology. . . . It is an interesting field and a fascinating one and one that no doubt holds problems for unraveling some of the mysteries in the biological organism; however, the field at present seems to be awaiting some new breakthrough in our understanding of biology."

Hopefully, I have been careful enough in my studies and in this book and have succeeded in avoiding the mythology. In my opinion, one breakthrough has come in the location of sites of pacemakers and elucidation of their function, and I expect another breakthrough to be forthcoming in the discovery of the cellular mechanisms. In this book, however, I have succeeded if I have increased your interest in the time of your body.

APPENDIX

MAKE YOUR OWN CIRCADIAN RHYTHM CHART

Make a photocopy of the chart. Time of day is on the horizontal axis (times at top of chart; day of month is on the vertical axis). To make your record, simply blacken the squares for each hour you are awake each day. If you travel across time zones, or change from daylight savings to standard time, during your record-keeping, put the times of the new time zone across the bottom axis and use them to guide you in plotting your waking hours (your alpha!). If you wish to continue, just make a chart for the next month; to see your entire record, lay the charts vertically in chronological order; if you make a copy of your chart you can make a double plot (as shown in Fig. 1.11).

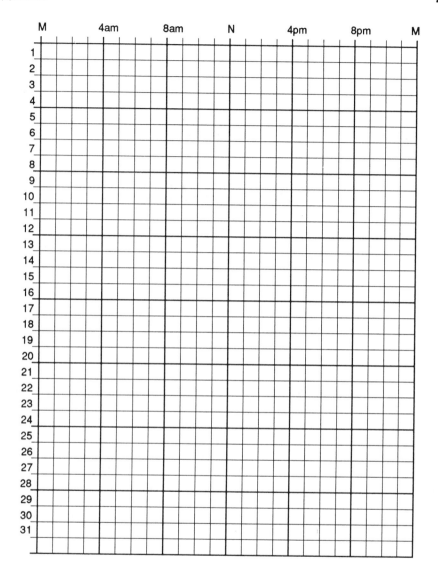

REFERENCES

References on biological clocks are distributed among the many journals in biology, psychiatry, and medicine. Three journals now specialize in articles of the field of biological clocks. The *Journal of Interdisciplinary Cycle Research* began publication in 1970, followed by *Chronobiology* in 1973. Growth of the field is reflected by the addition of yet a third journal, the *Journal of Biological Rhythms*, in 1986.

The references in the list here are not exhaustive; they represent but a tiny portion of the available literature. However, the individual bibliographies published in the references in this list, plus the lists of references in the books in the Annotated Bibliography, added to the references cited in the three journals, should lead to many of the publications in the field.

[1] Alleva, J. "The biological clock and pineal gland: How they control seasonal fertility in the golden hamster," *Pineal Research Reviews*, 5 (1987), 95–131.

[2] Alleva, J., M. Waleski, and F. Alleva, "A biological clock controlling the estrous cycle of the hamster" *Endocrinology*, 88 (1971), 1368–79.

[3] *An atlas of phase response curves for circadian pacemakers* (Clock Workshop; Hopkins Marine Station, 1977).

[4] Anderson, R., D. Laval-Martin, and L. Edmunds, "Cell cycle oscillators: Temperature compensation of the circadian rhythm of cell division in *Euglena*," *Experimental Cell Research*, 157: (1985), 144–58.

[5] Andrews, R. "Circadian rhythms in adrenal organ cultures," *Gegenbaurs Morph. Jahrb. Leipzig*, 117: (1971), 89–98.

[6] Arendt, J. "Light and melatonin as Zeitgebers in man," *Chronobiology International*, 4 (1987), 273–82.

[7] Arendt, J., M. Aldhous, and V. Marks, "Alleviation of jet lag by melatonin: preliminary results of controlled double blind trial," *British Medical Journal*, 292 (1986), 1170.

[8] Aschoff, J. "Tierische Periodik unter dem Einfluss von *Zeitgebern, Z. Tierpsychol*, 15 (1958), pp 1–30.

[9] Aschoff, J. ed., *Circadian Clocks* (Amsterdam: North-Holland Pub. Co., 1965), 479 pages.

[10] Aschoff, J. "Response curves in circadian periodicity" in *Circadian Clocks* ed. J. Aschoff (Amsterdam: North-Holland Pub. Co., 1965), pp. 95–111.

[11] Aschoff, J. "The phase-angle difference in circadian periodicity," in *Circadian Clocks* ed. J. Aschoff (Amsterdam: North-Holland Pub. Co., 1965), pp. 262–76.

[12] Aschoff, J. "Circadian activity rhythms in chaffinches (*Fringilla coelebs*) under constant conditions" *The Japanese Journal of Physiology*, 16 (1966), 363–70.

[13] Aschoff, J. "Circadian rhythms: General features and endocrinological aspects," in *Endocrine Rhythms*, ed. D. Krieger (New York: Raven Press, 1979), pp. 1–61.

[14] Aschoff, J. *Biological Rhythms*, vol. 4 of *Handbook of Behavioral Biology*, (New York: Plenum Press, 1981), 562 pages.

[15] Aschoff, J. "A survey on biological rhythms," in *Biological Rhythms*, ed. J. Aschoff (New York: Plenum Press, 1981), pp. 3–10.

[16] Aschoff, J. "Freerunning and entrained circadian rhythms," in *Biological Rhythms*, ed. J. Aschoff (New York: Plenum Press, 1981), pp. 81–93.

[17] Aschoff, J., S. Daan, and G. A. Groos, eds., *Vertebrate Circadian Systems: Structure and Physiology* (New York: Springer-Verlag New York, Inc., 1982), 363 pages.

[18] Aschoff, J., M. Fatranska, and H. Giedke, "Human circadian rhythms in continuous darkness: Entrainment by social cues," *Science*, 171 (1971), 213–15.

[19] Aschoff, J., K. Hoffman, H. Pohl, and R. Wever, "Re-entrainment of circadian rhythms after phase-shifts of the Zeitgeber," *Chronobiologia*, 2 (1975), 23–78.

[20] Aschoff, J. and C. Von Goetz, "Masking of circadian activity rhythms in hamsters by darkness," *J. Comp. Physiol. A.*, 162 (1988), 559–62.

[21] Aschoff, J. and R. Wever, "The circadian system of man," in *Biological Rhythms*, ed. J. Aschoff (New York: Plenum Press, 1981), pp. 311–31.

[22] Austin, C. and R. Short, *Hormones in Reproduction* (New York: Cambridge University Press, 1972), p. 69.

[23] Barfuss, D. and L. Ellis, "Seasonal cycles in melatonin synthesis by the pineal gland as related to testicular function in the house sparrow," *Gen. Comp. Endocrin.*, 17 (1971), 183–93.

[24] Bargiello, T. and M. Young, "Molecular genetics of a biological clock in *Drosophila*," *Proc. Natl. Acad. Sci. USA*, 81 (1984), 2142–46.

[25] Barnes, R. *Invertebrate Zoology* (Philadelphia: Saunders, 1963), p. 9.

[26] Barnwell, F. "Daily and tidal patterns of activity in individual fiddler crabs (Genus *Uca*) from the Woods Hole region," *Biol. Bull.*, 130 (1966), 1–17.

[27] Benoit, J. and L. Ott, "External and internal factors in sexual activity. Effect of irradiation with different wavelengths on the mechanisms of photostimulation of the hypophysis and on testicular growth in the immature duck," *Yale J. of Biol. Med.*, 17 (1944), 22–46.

[28] Besharse, J. "The daily light-dark cycle and rhythmic metabolism in the photoreceptor-pigment epithelial complex," in *Progress in Retinal Research*, eds. N. Osborne and G. Chader, (New York: Pergamon Press, Inc., 1982), pp. 81–124.

[29] Besharse, J. and P. Iuvone, "Circadian clock in *Xenopus* eye controlling retinal serotonin N-acetyltransferase," *Nature*, 305 (1983), 133–35.

[30] Binkley, S. *Circadian organization in the house sparrow, Passer domesticus*. Ph.D diss., Univ. of Texas at Austin, (1971), 168 pages.

[31] Binkley, S. "Rhythm analysis of clipped data: Examples using circadian data," *J. Comp. Physiol.*, 85 (1973), 141–46.

[32] Binkley, S. "Pineal and melatonin: circadian rhythms and body temperature," in *Chronobiology*, eds. L. Scheving, F. Halberg, and J. Pauly (Tokyo: Igaku Shoin, 1974), pp. 582–85.

[33] Binkley, S. "Comparative pineal biochemistry in birds and mammals," *American Zoologist*, 16 (1976), 57–65.

[34] Binkley, S. "Computer methods of analysis of biorhythm data," in *Biorhythms in the Marine Environment*, ed. P. DeCoursey (South Carolina: University of South Carolina Press, 1976), pp. 53–62.

[35] Binkley, S. "Pineal gland biorhythms: N-acetyltransferase in chickens and rats," *Federation Proceedings*, 35 (1976), 2347–52.

[36] Binkley, S. "Constant light: Effects on the circadian locomotor rhythm in the house sparrow," *Phys. Zoo.*, 50 (1977), 170–81.

[37] Binkley, S. "Light-to-dark transitions and dark-time sensitivity: Importance for the biological clock in the house sparrow," *Phys. Zoo.*, 51 (1978), 272–78.

[38] Binkley, S. "A timekeeping enzyme in the pineal gland," *Scientific American*, 24 (1979), 50–55.

[39] Binkley, S. "Circadian rhythm in pineal N-acetyltransferase activity: Phase shifting by light pulses II," *J. Neurochemistry*, 41 (1983), 273–76.

[40] Binkley, S. "Circadian rhythms of pineal function in rats," *Endocrine Reviews*, 4 (1983), 255–70.

[41] Binkley, S. "Rhythms in ocular and pineal N-acetyltransferase: A portrait of an enzyme clock," *Comp. Biochem. Physiol.*, 75A (1983), 123–29.

[42] Binkley, S. "Melatonin and N-acetyltransferase rhythms in pineal and retina," in *Pineal and Retinal Relationships*, eds. D. Klein and P. O'Brien (New York: Pergamon Press, Inc., 1986), pp. 185–96.

[43] Binkley, S. "Circadian locomotor rhythms in birds correlate with and may be explained by rhythms in serotonin, N-acetyltransferase, and melatonin," in *Processing of Environmental Information in Vertebrates*, ed. M. Stetson (New York: Springer-Verlag New York, Inc., 1988), pp. 85–100.

[44] Binkley, S. *The Pineal: Endocrine and Nonendocrine Function* (Englewood Cliffs, N.J.: Prentice-Hall, Inc., 1988), 304 pages.

[45] Binkley, S. and E. Geller, "Pineal enzymes in chickens: Development of daily rhythmicity," *Gen. and Comp. Endo.*, 27 (1975), 424–29.

[46] Binkley, S. and E. Geller, "Pineal N-acetyltransferase in chickens: Rhythm persists in constant darkness," *J. Comp. Physiol.*, 99 (1975), 67–70.

[47] Binkley, S., M. Hryshchyshyn, and K. Reilly, "N-acetyltransferase activity responds to environmental lighting in the eye as well as the pineal gland," *Nature*, 281 (1979), 479–81.

[48] Binkley, S., D. Klein, and J. Weller, "Dark initiation of the daily increase in pineal serotonin N-acetyltransferase," *Experientia*, 29 (1973), 1339.

[49] Binkley, S., S. Klein, and K. Mosher, "Light and dark control circadian phase in sparrows," in *The Pineal Gland, Endocrine Aspects*, eds. G. Brown and S. Wainwright, (New York: Pergamon Press, Inc., 1984), pp. 59–65.

[50] Binkley, S., E. Kluth, and M. Menaker, "Pineal function in sparrows: circadian rhythms and body temperature," *Science*, 174 (1971), 311–14.

[51] Binkley, S., E. Kluth, and M. Menaker, "Pineal and locomotor activity: Levels and arrhythmia in sparrows," *J. Comp. Physiol.*, 77 (1972), 163–69.

[52] Binkley, S. and K. Mosher, "Circadian rhythm in pineal N-acetyltransferase activity III: Phase shifting by dark pulses," *J. Neurochem.*, 45 (1985), 875–78.

[53] Binkley, S. and K. Mosher, "Direct and circadian control of perching behavior in sparrows," *Physiol. Behav.*, 35 (1985), 785–97.

[54] Binkley, S. and K. Mosher, "Oral melatonin produces arrhythmia in sparrows," *Experientia*, 41 (1985), 1615–17.

[55] Binkley, S. and K. Mosher, "Photoperiod modifies circadian resetting responses in sparrows," *American Journal of Physiology*, 251 (1986), R1156–62.

[56] Binkley, S. and K. Mosher, "Circadian perching in sparrows: Early responses to two light pulses," *Journal of Biological Rhythms*, 2 (1987), 1–11.

[57] Binkley, S. and K. Mosher, "Sparrow circadian rhythm responses to rotated light-dark schedules," *Physiology and Behavior*, 41 (1987), 361–70.

[58] Binkley, S. and K. Mosher, "Two circadian rhythms in pairs of sparrows," *J. Biological Rhythms*, 3 (1988), 249–54.

[59] Binkley, S., K. Mosher, and K. Reilly, "Circadian rhythms in house sparrows: Lighting ad lib," *Physiol. Behav.*, 31 (1983), 829–83.

[60] Binkley, S., K. Mosher, and B. White, "Circadian rhythm in pineal N-acetyltransferase activity: Rapid phase reversal and response to shorter than 24-hour cycles (IV)," *J. Neurochem.*, 49 (1987), 828–33.

[61] Binkley, S., D. Crawford, B. G. Tomé, and K. Mosher, "Self-reported sleep-wake schedules: A preliminary field study of almost feral humans," in preparation (1989).

[62] Binkley, S., S. MacBride, D. Klein, and C. Ralph, "Pineal enzymes: regulation of avian melatonin synthesis," *Science*, 181 (1973), 273–75.

[63] Binkley, S. and K. Mosher, "Advancing schedules and constant light produce faster resynchronization of circadian rhythms," Abstract 67, Society for Research on Biological Rhythms, 1988.

[64] Binkley, S., G. Muller, and T. Hernandez, "Circadian rhythm in pineal N-acetyltransferase activity: Phase shifting by light pulses I," *J. Neurochemistry*, 37 (1981), 798–800.

[65] Binkley, S., K. Reilly, V. Hermida, and K. Mosher, "Circadian rhythms of color change in *Anolis carolinensis:* Reconsideration of regulation, especially the role of melatonin in dark-time pallor," *Pineal Research Reviews*, 5 (1987), 133–51.

[66] Binkley, S., K. Reilly, and M. Hryshchyshyn, "N-acetyltransferase in the chick retina I: Circadian rhythms controlled by environmental lighting are similar to those in the pineal gland," *J. Comp. Physol.*, 139 (1980), 103–08.

[67] Binkley, S., J. Riebman, and K. Reilly, "Timekeeping by the pineal gland," *Science*, 197 (1977), 1181–83.

[68] Binkley, S., J. Riebman, and K. Reilly, "Regulation of pineal rhythms in chickens: Inhibition of the dark-time rise in N-acetyltransferase activity," *Comp. Biochem. Physiol.*, 59C (1978), 165–71.

[69] Binkley, S., J. Riebman, and K. Reilly, "The pineal gland: A biological clock *in vitro*," *Science*, 202 (1978), 1198–1201.

[70] Binkley, S., J. Riebman, and K. Reilly, "Regulation of pineal rhythms in chickens: N-acetyltransferase activity in homogenates," *Comp. Biochem. Physiol.*, 63C (1979), 291–96.

[71] Binkley, S., J. Stephens, J. Riebman, and K. Reilly. "Regulation of pineal rhythms in chickens: Photoperiod and dark-time sensitivity," *Gen. and Comp. Endo.*, 32 (1977), 411–16.

[72] Blackman, R. and J. Tukey, *The measurement of power spectra* (New York: Dover Publications, Inc., 1958), 190 pages.

[73] Block, G. and P. Davenport, "Circadian rhythmicity in *Bulla gouldiana:* Role of the eyes in controlling locomotor behavior," *J. Exp. Zoo.*, 224 (1982), 57–63.

[74] Block, G., D. McMahon, S. Wallace, and W. Friesen, "Cellular analysis of the *Bulla* ocular circadian pacemaker system. I. A model for retinal organization," *J. Comp. Physiol.*, A. 155 (1984), 365–78.

[75] Block, G. and S. Wallace, "Localization of a circadian pacemaker in the eye of a mollusc, *Bulla*," *Science*, 217 (1982), 155–57.

[76] Boulos, Z. and B. Rusak, "Circadian phase response curves for dark pulses in the hamster," *J. Comp. Physiol.*, 146 (1982), 411–17.

[77] Boyce, P. and D. Kennaway, "Effects of light on melatonin production," *Biol. Psychiatry*, 22 (1987), 473–78.

[78] Boyce, P. and G. Parker, "Seasonal affective disorder in the southern hemisphere," *Am. J. Psychiatry*, 145 (1988), 96–99.

[79] Brady, J. *Biological Timekeeping* (New York: Cambridge University Press, 1982), 197 pages.

[80] Brady, J. "Circadian rhythms—endogenous or exogenous? *J. Comp. Physiol.*, A 161 (1987), 711–14.

[81] Brammer, M., S. Binkley, and K. Mosher, "The rise and fall of pineal N-acetyltransferase *in vitro*: Neural regulation in the developing rat," *J. Neurobiology*, 13 (1982), 487–94.

[82] Broadway, J. and J. Arendt, "Delayed recovery of sleep and melatonin rhythms after nightshift work in antarctic winter," *The Lancet*, Oct. 4 (1985), 813–14.

[83] Broda, H., D. Brugge, K. Homma, and J. Hastings, "Circadian communication between unicells?," *Cell Biophysics*, 8 (1985), 47–67.

[84] Brown, F. "Response to pervasive geophysical factors and the biological clock problem," *Cold Spring Harbor Symposia on Quantitative Biology*, 25 (1960), 57–72.

[85] Brown, F., J. W. Hastings, and J. D. Palmer, *The Biological Clock, Two Views* (New York: Academic Press, Inc., 1970), 94 pages.

[86] Bruce, V. "Mutants of the biological clock in *Chlamydomonas reinhardi*," *Genetics*, 70 (1972), 537–48.

[87] Bünning, E. "Zur kenntnis der endonomen tagesrhythmik bei insekten und bei pflanzen," *Ber. Dtsch. Bot. Ges.*, 53 (1935), 594–623.

[88] Bünning, E. "Circadian rhythms and the time measurement in photoperiodism," *Cold Spring Harbor Symposia on Quantitative Biology*, 25 (1960), 249–56.

[89] Bünning, E. Opening Address: "Biological clocks," *Cold Spring Harbor Symposia on Quantitative Biology*, 25 (1960), pp. 1–9.

[90] Bünning, E. *The Physiologcal Clock*, 3rd ed. (New York: Springer Verlag New York, Inc., 1973), 167 pages.

[91] Campbell, C. and N. B. Schwartz, "The impact of constant light on the estrous cycle of the rat," *Endocrinology*, 106 (1980), 1230–38.

[92] Carmichael, M. and I. Zucker, "Circannual rhythms of ground squirrels: A test of the frequency demultiplication hypothesis," *J. Biological Rhythms*, 1 (1986), 277–84.

[93] Carter, D. and B. Goldman, "Antigonadal effects of timed melatonin infusion in pinealectomized male Djungarian hamsters (*Phodopus sungorus sungorus*): Duration is the critical parameter," *Endocrinology*, 113 (1983), 1261–73.

[94] Cavallini, M., F. Halberg, G. Cornelissen, F. Enrichens, and C. Margarit, "Organ transplantation and broader chronotherapy with implantable pump and computer programs for marker rhythm assessment," *Journal of Controlled Release*, 3 (1986), 3–13.

[95] Chandrashekaran, M. "Studies on phase-shifts of endogenous rhythms in *Drosophila pseudoobscura*," *Zeitschrift für Vergleichende Physiologie*, 56 (1967), 154–62.

[96] Chandrashekaran, M. "Studies on phase-shifts in endogenous rhythms. II. The dual effect of light on the entrainment of the eclosion rhythm in *Drosophila pseudoobscura*," *Zeitschrift für Vergleichende Physiologie* 56 (1967), 163–70.

[97] Chandrashekaran, M. "Apparent absence of a separate B-oscillator in phasing the circadian rhythm of eclosion of *Drosophila pseudoobscura*," in *Development and Neurobiology of Drosophila*, eds. O. Siddiqi, P. Babu, L. Hall, and J. Hall (New York: Plenum Pub. Corp., 1980), pp. 417–24.

[98] Chandrashekaran, M. "The *Drosophila* circadian clock," *Proc. Indian Acad. Sci. (Anim. Sci.)*, 94 (1985), 187–96.

[99] Chandrashekaran, M., A. Johnsson, and W. Engelmann, "Possible *dawn* and *dusk* roles of light pulses shifting the phase of a circadian rhythm," *J. Comp. Physiol.*, 82 (1973), 347–56.

[100] Chandrashekaran, M. and W. Engelmann. "Amplitude attenuation of the circadian rhythm in *Drosophila* with light pulses of varying irradiance and duration," *International Journal of Chronobiology*, 3 (1976), 231–40.

[101] Chesworth, M., V. Cassone, and S. Armstrong, "Effects of daily melatonin injections on activity rhythms of rats in constant light," *Am. J. Physiol.*, 253 (1987), R101–07.

[102] Claustrat, B., G. Chazot, J. Brun, D. Jordan, and G. Sassolas, "A chronobiological study of melatonin and cortisol secretion in depressed subjects: Plasma melatonin, a biochemical marker in major depression," *Biological Psychiatry*, 19 (1984), 1215–28.

[103] Cole, L. "Biological clock in the unicorn," *Science* 125 (1957), 874–76.

[104] Coleman, R. *Wide Awake at 3:00 A.M. By Choice or by Chance?* (New York: W. H. Freeman and Co., 1986), 195 pages.

[105] Conroy, R. and J. Mills, *Human Circadian Rhythms* (London: J & A Churchill, 1970), 236 pages.

[106] Corrent, G., D. McAdoo, and A. Eskin, "Serotonin shifts the phase of the circadian rhythm from the *Aplysia* eye," *Science*, 202 (1978), 977–79.

[107] Czeisler, C., J. Allan, S. Strogatz, J. Ronda, R. Sanchez, C. Rios, W. Freitag, G. Richardson, and R. Kronauer, "Bright light resets the human circadian pacemaker independent of the timing of the sleep-wake cycle," *Science*, 233 (1986), 667–71.

[108] Czeisler, C., M. Moore-Ede, and R. Coleman, "Rotating shift work schedules that disrupt sleep are improved by applying circadian principles," *Science*, 217 (1982), 460–63.

[109] Czeisler, C., G. Richardson, J. Zimmerman, M. Moore-Ede, and E. Weitzman, "Entrainment of human circadian rhythms by light-dark cycles: A reassessment," *Photochemistry and Photobiology*, 34 (1981), 239–47.

[110] Daan, S. and J. Aschoff, "Circadian contributions to survival." In *Vertebrate Circadian Systems*, eds. J. Aschoff, S. Daan, G. Groos, (New York: Springer-Verlag New York, Inc., 1982), pp. 305–21.

[111] Daan, S. and C. Pittendrigh, "A functional analysis of circadian pacemakers in nocturnal rodents. III. Heavy water and constant light: Homeostasis of frequency," *J. Comp. Physiol.*, 106 (1976), 267–90.

[112] Daan, S. and C. Pittendrigh, "A functional analysis of circadian pacemakers in nocturnal rodents. II. The variability of phase response curves," *J. Comp. Physiol.*, 106 (1976), 253–66.

[113] Daan, S., D. Damassa, C. Pittendrigh, and E. Smith, "An effect of castration and testosterone replacement on a circadian pacemaker in mice (*Mus musculus*)," *Proc. Nat. Acad. Sci. USA*, 72 (1975), 3744–47.

[114] Dark, J., G. Pickard, and I. Zucker, "Persistence of circannual rhythms in ground squirrels with lesions of the suprachiasmatic nuclei," *Brain Research*, 332 (1985), 201–7.

[115] Darwin, C. and F. Darwin, *The Power of Movement in Plants* (New York: D. Appleton and Co., 1881).

[116] Davis, F. "Development of the mouse circadian pacemaker: Independence from environmental cycles," *J. Comp. Physiol.*, 143 (1981), 527–39.

[117] Davis, F. "Ontogeny of circadian rhythms," in *Biological Rhythms*, ed. J. Aschoff, *Handbook of Behavioral Neurobiology*, 4 (1981), 257–74.

[118] Davis, F., S. Stice, and M. Menaker, "Activity and reproductive state in the hamster: Independent control by social stimuli and a circadian pacemaker," *Physiol. Behav.*, 40 (1987), 583–90.

[119] DeCoursey, P. *Daily Activity Rhythms in the Flying Squirrel, Glaucomys volans*. Ph.D. dissertation, Univ. of Wisconsin, 1959.

[120] DeCoursey, P. "Phase control of activity in a rodent." *Cold Spring Harbor Symposia on Quantitative Biology*, 25 (1960), 49–56.

[121] DeCoursey, P. "Function of a light response rhythm in hamsters," *J. Cell. Comp. Physiol.*, 63 (1964), 189–96.

[122] DeCoursey, P. "Circadian photoentrainment: Parameters of phase delaying," *Journal of Biological Rhythms*, 3 (1986), 171–86.

[123] Deguchi, T. "A circadian oscillator in cultured cells of chicken pineal gland," *Nature*, 282 (1979), 94–96.

[124] Deguchi, T. "Ontogenesis and phylogenesis of circadian rhythm of serotonin N-acetyltransferase activity in the pineal gland," in *Biological Rhythms and Their Central Mechanism*, eds. M. Suda, O. Hayaishi, and H. Nakagawa (Amsterdam: Elsevier/North-Holland Pub. Co., 1979), pp. 159–68.

[125] Dewan, E. "On the possibility of a perfect rhythm method of birth control by periodic light stimulation," *American Journal of Obstetrics and Gynecology*, 99 (1967), 1016–19.

[126] Dunlap, J., W. Taylor, and J. Hastings, "The effects of protein synthesis inhibitors on the *Gonyaulax* clock. I. Phase-shifting effects of cycloheximide," *J. Comp. Physiol.*, 138 (1980), 1–8.

[127] Earnest, D. and F. Turek, "Splitting of the circadian rhythm of activity in hamsters: Effects of exposure to constant darkness and subsequent re-exposure to constant light," *J. Comp. Physiol.*, 145 (1982), 405–11.

[128] Ebihara, S. "Circadian organization in the pigeon, *Columba livia:* The role of the pineal organ and the eye," *J. Comp. Phys.*, 154 (1984), 59–69.

[129] Ebihara, S. and H. Kawamura, "The role of the pineal organ and the suprachiasmatic nucleus in the control of circadian locomotor rhythms in the Java sparrow, *Padda orzivora*," *J. Comp. Physiol.*, 141 (1981), 207–14.

[130] Edmonds, S. and N. Adler, "Food and light as entrainers of circadian running activity in the rat," *Physiol. Behav.*, 18 (1977), 915–19.

[131] Edmonds, S. and N. Adler, "The multiplicity of biological oscillators in the control of circadian running activity in the rat." *Physiol. Behav.*, 18 (1977), 921–30.

[132] Edmunds, L. "Replication of DNA and cell division in synchronously dividing cultures of *Euglena gracilis*," *Science*, 145 (1964), 266–68.

[133] Edmunds, L. "Studies on synchronously dividing cultures of *Euglena gracilis Klebs* (Strain Z)," *J. Cell. and Comp. Physiol.*, 66 (1966), 147–58.

[134] Edmunds, L. "Persistent circadian rhythm of cell division in *Euglena:* Some theoretical considerations and the problem of intercellular communication," in *Biochronometry*, ed. M. Menaker (Washington: National Academy of Sciences, 1971), pp. 594–611.

[135] Edmunds, L. "Temporal differentiation in *Euglena:* Circadian phenomena in non-dividing populations and in synchronously dividing cells," in *Les cycles cellulaires et leur blocage chez plusieurs protistes*, Colloques Internationaux du Centre National de la Recherche Scientifique 240 (Paris: Centre National de la Recherche Scientifique, 1976), pp. 53–67.

[136] Edmunds, L. and F. Halberg, "Circadian time structure of *Euglena:* A model system amenable to quantification," in *Neoplasms*, ed. H. Kaiser (Baltimore: The Williams and Wilkins Company, 1981), pp. 105–34.

[137] Edmunds, L., M. Jay, A. Kohlmann, S. Liu, V. Merriam, and H. Sternberg, "The coupling effects of some thiol and other sulfur containing compounds on the circadian rhythm of cell division in photosynthetic mutants of *Euglena*," *Arch. Microbiol*, 108 (1976), 1–8.

[138] Edmunds, L. and D. Laval-Martin, "Cell division cycles and circadian oscillators," in *Cell Cycle Clocks*, ed. L. Edmunds (New York: Marcel Dekker, Inc., 1984), pp. 295–324.

[139] Edmunds, L., D. Tay, and D. Laval-Martin, "Cell division cycles and circadian clocks: Phase-response curves for light perturbations in synchronous cultures of *Euglena*," *Plant Physiol.*, 70 (1982), 297–302.

[140] Ehret, C. F. and E. Trucco, "Molecular models for the circadian clock. I. The chronon concept," *J. Theoret. Biol.*, 15 (1967), 240–62.

[141] Ehret, C. F. and L. W. Scanlon, *Overcoming Jet Lag* (New York: Berkley Pub. Corp. 1983), 160 pages.

[142] Elliott, J. "Circadian rhythms and photoperiodic time measurement in mammals," *Federation Proceedings*, 35 (1976), 2339–46.

[143] Elliott, J., M. Stetson, and M. Menaker, "Regulation of testis function in golden hamsters: A circadian clock measures photoperiodic time," *Science*, 178 (1972), 771–73.

[144] Ellis, G., R. McKlveen, and F. Turek, "Dark pulses affect the circadian rhythm of activity in hamsters kept in constant light," *Am. J. Physiol.*, 242 (1982), R44–50.

[145] Emlen, S. "Migration: Orientation and navigation," in *Avian Biology*, V, eds. D. Farner and J. King (New York: Academic Press, Inc., 1975), pp. 129–219.

[146] Enright, J. "The internal clock of drunken isopods. *Z. Vergl. Physiologie*, 75 (1971), 332–46.

[147] Enright, J. "The search for rhythmicity in biological time series," *J. Theoretical Biology*, 8 (1957), 426–68.

[148] Enright, J. *The timing of sleep and wakefulness: On the substructure and dynamics of the circadian pacemakers underlying the wake-sleep cycle* (New York: Springer-Verlag New York, Inc., 1980), 263 pages.

[149] Enright, J. and A. Winfree, "Detecting a phase singularity in a coupled stochastic system," *Lectures on Mathematics in the Life Sciences*, 19 (1987), 121–50.

[150] Eskin, A. "Some properties of the system controlling the circadian activity rhythm of sparrows," in *Biochronometry*, ed. M. Menaker (Washington, D.C.: National Academy of Sciences, 1971), pp. 55–77.

[151] Eskin, A. and J. Takahashi, "Adenylate cyclase activation shifts the phase of a circadian pacemaker," *Science*, 220 (1983), 82–84.

[152] Eskin, A., G. Corrent, C. Lin, and D. McAdoo, "Mechanism for shifting the phase of a circadian rhythm by serotonin: Involvement of cAMP," *Proc. Natl. Acad. Sci. U.S.A.*, 79 (1982), 660–64.

[153] Eskin, A., J. Takahashi, M. Zatz, and G. Block, "Cyclic guanosine 3':5'-monophosphate mimics the effects of light on a circadian pacemaker in the eye of *Aplysia*," *The Journal of Neuroscience*, 4 (1984), 2466–71.

[154] Feldman, J. "Genetics of circadian clocks," *BioScience*, 33 (1983), 426–31.

[155] Feldman, J. and M. Hoyle, "Complementation analysis of linked circadian clock mutants of *Neurospora crassa*," *Genetics*, 82 (1976), 9–17.

[156] Finocchiaro, L., J. Callebert, J. Launay, and J. Jallon, "Melatonin biosynthesis in *Drosophila*: Its nature and effects," *Journal of Neurochemistry*, 50 (1988), 382–87.

[157] Fletcher, C. *The Man Who Walked Through Time* (New York: Random House Vintage Books, 1967), p. 214.

[158] Follett, B. and D. Follett, *Biological Clocks in Seasonal Reproductive Cycles* (New York: John Wiley & Sons, Inc., 1981), 292 pages.

[159] Frank, K. and W. Zimmerman, "Action spectra for phase shifts of a circadian rhythm in *Drosophila*," *Science*, (1969), 688–89.

[160] Franklin, B. *Poor Richard's Almanac* (Philadelphia: Printed and sold by Benjamin Franklin).

[161] Ganong, W. *Review of Medical Physiology*. Norwalk, CT: Appleton & Lange, 1987), p. 450.

[162] Gaston, S. and M. Menaker, "Pineal function: The biological clock in the sparrow?" *Science*, 160 (1968), 1125–27.

[163] Gilbert, D. A. (poem, no title), in *Cell Cycle Clocks*, ed. L. Edmunds. (New York: Marcel Dekker, Inc., 1984), p. 2.

[164] Goto, K., D. Laval-Martin, and L. Edmunds, "Biochemical modeling of an autonomously oscillatory circadian clock in *Euglena*," *Science*, 228 (1985), 1284–88.

[165] Gwinner, E. *Circannual Rhythms: Endogenous Annual clocks in the Organizatio of Seasonal Processes* (New York: Springer-Verlag New York, Inc., 1986), 15 pages.

[166] Hadley, M. *Endocrinology* (Englewood Cliffs, NJ: Prentice-Hall, Inc., 1988), ￼ 440.

[167] Halberg, F. "Temporal coordination of physiologic function," *Cold Spring Harbe Symposium*, 25 (1960), 289–310.

[168] Halberg, F. *Glossary of Chronobiology* (Milano, Italia: Il Ponte, 1977), 188 page:

[169] Halberg, F., E. Halberg, C. Barnum, and J. Bittner, "Physiologic 24-hour per odicity in human beings and mice, the lighting regimen and daily routine," i *Photoperiodism and Related Phenomena in Plants and Animals*, ed. R. Withrov Ed. Publ. No. 55 of the Amer. Assoc. Adv. Sci. (Washington, D.C., 1959), p￼ 803–78.

[170] Halberg, F., Y. L. Tong, and E. A. Johnson, "Circadian System Phase—An aspe of temporal morphology; Procedures and illustrative examples," in *The Cellula Aspects of Biorhythms*, Symposium on Biorhythms (New York: Springer-Verla New York, Inc., 1967), pp. 20–48.

[171] Hamblen, M., W. Zehring, C. Kyriacou, P. Reddy, Q. Yu, D. Wheeler, L. Zwiebe R. Konopka, M. Rosbash, J. Hall, "Germ-line transformation involving DN from the period locus in *Drosophila melanogaster:* Overlapping genomic fra; ments that restore circadian and ultradian rhythmicity to the per^p and per mutants," *J. Neurogenetics*, 3 (1986), 249–91.

[172] Haman, J. "The length of the menstrual cycle. A study of 50 normal women," *An J. Obstet. Gynecol.*, 43 (1942), 870–73.

[173] Hamner, K. C. "Photoperiodism and circadian rhythms," *Cold Spring Harbe Symposium on Quantitative Biology*, 25 (1960), 269–77.

[174] Handler, A. and R. Konopka, "Transplantation of a circadian pacemaker ￼ *Drosophila*," *Nature*, 279 (1979), 235–38.

[175] Hastings, J. "Biochemical aspects of rhythms: Phase shifting by chemicals *Cold Spring Harbor Symposium on Quantitative Biology*, 25 (1960), 131–43.

[176] Hawking, S. *A Brief History of Time* (New York: Bantam Books, Inc., 1988), 1￼ pages.

[177] Hoffman, K. "Zum einfluss der zeitgeberstarke auf die phasenlage der sy chronisierten circadianen periodik," *Z. Vergl. Physiologie*, 62 (1969), 93–110.

[178] Hoffman, R. and R. Reiter, "Pineal gland: Influence on gonads of male ham ters," *Science*, 148 (1965), 1609–11.

[179] Honma, K., S. Honma, and T. Hiroshige, "Response curve, free-running perio and activity time in circadian locomotor rhythm of rats," *Japanese Journal Physiology*, 35 (1985), 643–58.

[180] Honma, K., S. Honma, and T. Wada, "Phase-dependent shift of free-runnir human circadian rhythms in response to a single bright light pulse," *Experienti* 43 (1987), 1205–07.

[181] Honma, K., S. Honma, and T. Wada, "Entrainment of human circadian rhythm by artificial bright light cycles," *Experientia*, 43 (1987), 572–74.

[182] Honma, S., K. Honma, and T. Hiroshige, "Restricted daily feeding during nursing period resets circadian locomotor rhythm of infant rats," *Am. J. Physiol.*, 252 (1987), R262–68.

[183] Illnerová, H. and J. Vaněček, "Pineal rhythm in N-acetyltransferase activity in rats under different artificial photoperiods and in natural daylight in the course of a year," *Neuroendocrinology*, 31 (1980), 321–26.

[184] Illnerová, H. and J. Vaněček, "Dynamics of discrete entrainment of the circadian rhythm in the rat pineal N-acetyltransferase activity during transient cycles," *Journal of Biological Rhythms*, 2 (1987), 95–108.

[185] Illnerová, H., K. Hoffmann, and J. Vaněček, Adjustment of pineal melatonin and N-acetyltransferase rhythms to change from long to short photoperiod in the Djungarian hamster *Phodopus sungorus*," *Neuroendocrinology*, 38 (1984), 226–31.

[186] Illnerová, H., P. Zvolsky, and J. Vaněček, "The circadian rhythm in plasma melatonin concentration of the urbanised man: the effect of summer and winter time," *Brain Research*, 328 (1985), 186–89.

[187] Isaacs, G., D. Stainer, T. Sensky, S. Moor, and C. Thompson, "Phototherapy and its mechanisms of action in seasonal affective disorder," *Journal of Affective Disorders*, 14 (1988), 13–19.

[188] Jacklet, J. "Circadian rhythm of optic nerve impulses recorded in darkness from isolated eye of *Aplysia*," *Science*, 164 (1969), 562–63.

[189] Jacklet, J. "Circadian locomotor activity in *Aplysia*," *J. Comp. Physiol.*, 79 (1972), 325–41.

[190] Jacklet, J. "Circadian rhythm from the eye of *Aplysia*: Temperature compensation of the effects of protein synthesis inhibitors," *J. Exp. Biol.*, 84 (1980), 1–15.

[191] Jacklet, J. "Neural organization and cellular mechanisms of circadian pacemakers," *International Review of Cytology*, 89 (1984), 251–94.

[192] Jouvet, M. "Neurophysiological and biochemical mechanisms of sleep," in *Sleep*, ed. A. Kales (Philadelphia, PA: J. B. Lippincott Co., 1969), pp. 89–100.

[193] Kales, A. *Sleep: Physiology and Pathology* (Philadelphia, PA: J. B. Lippincott Co., 1969), 360 pages.

[194] Kanabrocki, E., L. Scheving, F. Halberg, R. Brewer, and T. Bird, "*Circadian variation in presumably healthy young soldiers*," Document #PB 228427 (1969). National Technical Information Service, US Department of Commerce, P.O. Box 1553, Springfield, Virginia 22151.

[195] Keeton, W. *Biological Sciences*, 3rd ed. (New York: W. W. Norton & Co., 1980), 1080 pages.

[196] Kennaway, D., J. Peek, T. Gilmore, and P. Royles, "Pituitary response to LHRH, LH pulsatility and plasma melatonin and prolactin changes in ewe lambs treated with melatonin implants to delay puberty," *J. Reprod. Fert.*, 78 (1986), 137–48.

[197] Kennaway, D. and P. Royles, "Circadian rhythms of 6-sulphtoxy melatonin, cortisol and electrolyte excretion at the summer and winter solstices in normal men and women," *Acta Endocrinologica*, 113 (1986), 450–56.

[198] Klein, D. and J. Weller, "Indole metabolism in the pineal gland: A circadian rhythm in N-acetyltransferase," *Science*, 169 (1970), 1093–95.

[199] Klein, D. and J. Weller, "Rapid light-induced decrease in pineal serotonin N-acetyltransferase activity," *Science*, 177 (1972), 532–33.

[200] Klein, S., S. Binkley, and K. Mosher, "Circadian phase of sparrows: Control by light and dark," *Photochem. Photobio.*, 41 (1985), 453–57.

[201] Kleinhoonte, A. "Uber die durch das Licht regulierten autonomen Bewegungen der Canavalia-Blatter," *Archs. neerl. Sci., ser. IIIb*, 5 (1929), 1–110, 181.

[202] Kleitman, N. and T. Englemann, "Sleep characteristics of infants," *Journal of Applied Physiology*, 7 (1953), 269–82.

[203] Konopka, R. "Genetics and development of circadian rhythms in invertebrates," in *Biological Rhythms*, ed. J. Aschoff, *Handbook of Behavioral Biology vol. 4* (New York: Plenum Publishing Corp., 1981), pp. 173–81.

[204] Konopka, R. and S. Benzer, "Clock mutants of *Drosophila melanogaster*," *Proc. Nat. Acad. Sci. USA*, 68 (1971), 2112–16.

[205] Kraft, I., S. Alexander, D. Foster, R. Leachman, and H. Lipscomb, "Circadian rhythms in human heart homograft," *Science*, 169 (1970), 694–96.

[206] Kripke, D., S. Risch, and D. Janowsky, "Bright white light alleviates depression," *Psychiatry Research*, 10 (1983), 105–12.

[207] Kripke, D., D. Mullaney, M. Atkinson, and S. Wolf, "Circadian rhythm disorders in manic-depressives," *Biol. Psychiatry*, 13 (1975), 335–50.

[208] Laval-Martin, D., D. Shuch, and L. Edmunds, "Cell cycle-related and endogenously controlled circadian photosynthetic rhythms in *Euglena*," *Plant Physiol.* 63 (1979), 495–502.

[209] Lewy, A. "Effects of light on human melatonin production and the human circadian system," *Prog. Neuro-Psychopharmacol. & Biol. Psychiat.*, 7 (1983) 551–56.

[210] Lewy, A. and D. Newsome, "Different types of melatonin circadian secretory rhythms in some blind subjects," *J. Clin. Endocrinol. Metab.*, 56 (1983), 1103–07

[211] Lewy, A., T. Wehr, F. Goodwin, D. Newsome, and S. Markey, "Light suppresses melatonin secretion in humans," *Science*, 210 (1980), 1267–69.

[212] Lickey, M., J. Wozniak, G. Block, D. Hudson, and G. Augter, "The consequences of eye removal for the circadian rhythm of behavioral activity in *Aplysia*," *J. Comp Physiol.*, 118 (1977), 121–43.

[213] Livermore, A. and J. Stevens, "Light transducer for the biological clock: a function for rapid eye movements," *J. Neural Trans.*, 72 (1988), 37–42.

[214] Loher, W. "Circadian control of stridulation in the cricket, *Teleogryllus commodus* Walker," *J. Comp. Physiol.*, 79 (1972), 173–90.

[215] Luce, G. *Biological Rhythms in Psychiatry and Medicine*, Public Health Service Publication No. 2088 (U.S. Government Printing Office, 1970), 183 pages.

[216] Luce, G. *Body Time: Physiological Rhythms and Social Stress* (New York: Bantam Books, Inc., 1971), 441 pages.

[217] Lynch, H. J., R. Rivest, and R. Wurtman, "Artificial induction of melatonin rhythms by programmed microinfusion," *Neuroendocrinology*, 31 (1980) 106–11.

[218] MacBride, S. *Pineal Biochemical Rhythms of the Chicken (Gallus domesticus): Light Cycle and Locomotor Activity Correlates,*" Ph.D. thesis, University of Pittsburgh, Pittsburgh, PA, (1973), 209 pages.

[219] Magnus, G., M. Cavallini, F. Halberg, G. Cornelissen, D. Sutherland, J. Najarian, and W. Hrushesky, "Circadian toxicology of cyclosporin," *Toxicology and Applied Pharmacology,* 77 (1985), 181–85.

[220] Malinowski, J., D. Laval-Martin, and L. Edmunds, "Circadian oscillators, cell cycles, and singularities: light perturbations of the free-running rhythm of cell division in *Euglena,*" *J. Comp. Physiol. B.,* 155 (1985), 257–67.

[221] Matsumoto, K. and Y. Morita, "Effects of nighttime naps and age on sleep patterns of shift workers," *Sleep,* 10 (1987), 580–89.

[222] McClintock, M. "Menstrual synchrony and suppression," *Nature,* 229 (1971), 244–45.

[223] McDaniel, M., F. Sulzman, and J. Hastings, "Heavy water slows the *Gonyaulax* clock: A test of the hypothesis that D_2O affects circadian oscillations by diminishing the apparent temperature," *Proc. Nat. Acad. Sci. USA,* 71 (1974), 4389–91.

[224] McMillan, J., J. Elliott, and M. Menaker, "On the role of eyes and brain photoreceptors in the sparrow: Arrhythmicity in constant light," *J. Comp. Physiol.,* 102 (1975), 263–68.

[225] McMurray, L. and J. Hastings, "Circadian rhythms: Mechanism of luciferase activity changes in *Gonyaulax,*" *The Biological Bulletin,* 143 (1972), 196–206.

[226] Meier-Koll, A., U. Hall, U. Hellwig, G. Kott, and V. Meier-Koll, "A biological oscillator system and the development of sleep-waking behavior during early infancy," *Chronobiologia,* 5 (1978), 425–40.

[227] Meijer, J., E. van der Zee, and M. Dietz, "Glutamate phase shifts circadian activity rhythms in hamsters," *Neuroscience,* Letters 86 (1988), 177–83.

[228] Menaker, M. "Circadian rhythms and photoperiodism in *Passer domesticus,*" in *Circadian Clocks,* ed. J. Aschoff (Amsterdam: North-Holland Pub. Co., 1965), pp. 385–95.

[229] Menaker, M. "Light perception by extra-retinal receptors in the brain of the sparrow," *Proceedings, 76th Annual Convention, APA,* (1968), 299–300.

[230] Menaker, M. ed. *Biochronometry* (Washington, D.C., 1971), 662 pages.

[231] Menaker, M. and A. Eskin, "Entrainment of circadian rhythms by sound in *Passer domesticus,*" *Science,* 154 (1966), 1579–81.

[232] Menaker, M. and S. Wisner, "Temperature-compensated circadian clock in the pineal of *Anolis,*" *Proc. Nat. Acad. Sci. USA,* 80 (1983), 6119–21.

[233] Mercer, D. "The behavior of multiple oscillator systems," in *Circadian Clocks,* J. Aschoff, ed. (Amsterdam: North-Holland Pub. Co., 1965), pp. 64–73.

[234] Meyersbach, H., L. E. Scheving, and J. E. Pauly, eds., *Biological rhythms in structure and function* (New York: Alan R. Liss, 1981), 241 pages.

[235] Moore, R. and D. Klein, "Visual pathways and the central neural control of a circadian rhythm in pineal serotonin N-acetyltransferase activity," *Brain Research,* 71 (1974), 17–33.

[236] Moore-Ede, M. and C. Czeisler, *Mathematical Models of the Circadian Sleep-Wake Cycle* (New York: Raven Press, 1984), 216 pages.

[237] Moore-Ede, M. and G. Richardson, 1985. "Medical implications of shift-work," *Ann. Rev. Med.*, 36 (1985), 607–17.

[238] Moore-Ede, M., W. Schmelzer, D. Kass, and J. Herd, "Internal organization of the circadian timing system in multicellular animals," *Fed. Proc.*, 35 (1976), 2333–38.

[239] Moore-Ede, M., F. Sulzman, and C. Fuller, *The Clocks That Time Us: Physiology of the Circadian Timing System* (Cambridge, MA: Harvard University Press, 1982), 448 pages.

[240] Morin, L., K. Fitzgerald, and I. Zucker, "Estradiol shortens the period of hamster circadian rhythms," *Science*, 196 (1977), 305–07.

[241] Mrosovsky, N. "Phase response curves for social entrainment," *J. Comp. Physiol. A*, 162 (1988), 35–46.

[241b] Nishiitsutsuji-Uwo, J. and C. Pittendrigh. "Central nervous system control of circadian rhythmicity in the cockroach. III. The optic lobes, locus of the driving oscillation," *Z. Vergl. Physiol.*, 58 (1968), 14–46.

[242] Njus, D., F. Sulzman, and J. Hastings, "Membrane model for the circadian clock," *Nature*, 248 (1974), 116–20.

[243] Njus, D., V. Gooch, D. Mergenhagen, F. Sulzman, and J. Hastings. "Membranes and molecules in circadian systems," *Federation Proceedings*, 35 (1976), 2353–57.

[244] Nyce, J. and S. Binkley, "Extraretinal photoreception in chickens: Entrainment of the circadian locomotor activity rhythm," *Photochem. Photobiol.*, 25 (1977), 529–31.

[245] Page, T. "Transportation of the cockroach circadian pacemaker," *Science*, 216 (1982), 73–75.

[246] Page, T. "Circadian organization in cockroaches: Effects of temperature cycles on locomotor activity," *J. Insect. Physiol.*, 31 (1985), 235–42.

[247] Page, T. "Serotonin phase-shifts the circadian rhythm of locomotor activity in the cockroach," *J. Biological Rhythms*, 2 (1987), 23–34.

[248] Palmer, J. "Daily and tidal components in the persistent rhythmic activity of the crab, *Sesarma*," *Nature*, 215 (1967), 64–66.

[249] Palmer, J. *Biological Clocks in Marine Organisms* (New York: John Wiley & Sons, Inc., 1974), 173 pages.

[250] Palmer, J., F. Brown, and L. Edmunds *An Introduction to Biological Rhythms* (New York: Academic Press, 1976), 375 pages.

[251] Pasachoff, J. and M. Kutner *University Astronomy* (Philadelphia, PA: W. B. Saunders Co., 1978), 3665–74.

[252] Pavlidis, T. "Mathematical models of circadian rhythms: Their usefulness and their limitations," in *Biochronometry*, ed. M. Menaker (Washington, D.C.: National Academy of the Sciences, 1971), pp. 110–16.

[253] Pengelley, E. ed. *Circannual Clocks: Annual Biological Rhythms* (New York: Academic Press, Inc., 1974), 523 pages.

[254] Pengelley, E. and S. Asmundson, "Circannual rhythmicity in hibernating mammals," in *Circannual Clocks*, ed. E. Pengelley (New York: Academic Press, Inc., 1974), pp. 95–160.

[256] Pittendrigh, C. "Circadian rhythms and the circadian organization of living systems," *Cold Spring Harbor Symposia on Quantitative Biology*, XXV (1960), 159–84.

[257] Pittendrigh, C. "On temporal organization in living systems," *Harvey Lect.*, 56 (1961), 93–125.

[258] Pittendrigh, C. "On the mechanism of the entrainment of a circadian rhythm by light cycles," in *Circadian Clocks*, ed. J. Aschoff (Amsterdam: North-Holland, 1965), pp. 277–98.

[259] Pittendrigh, C. "The circadian oscillation in *Drosophila pseudoobscura* pupae: a model for the photoperiodic clock," *Z. Pflanzenphysiol.*, 54 (1966), 275–307.

[260] Pittendrigh, C. "Circadian systems: Entrainment," in *Biological Rhythms*, ed. J. Aschoff (New York: Plenum Publishing Corp., 1981), pp. 95–124.

[261] Pittendrigh, C. "Circadian systems: General perspective," in *Biological Rhythms*, ed. J. Aschoff (New York: Plenum Publishing Corp, 1981), pp. 57–80.

[262] Pittendrigh, C. and V. Bruce, "An oscillator model for biological clocks," in *Rhythmic and Synthetic Processes in Growth*, ed. D. Rudnick (Princeton, NJ: Princeton University Press, 1957), pp. 75–109.

[263] Pittendrigh, C. and P. Caldarola, "General homeostasis of the frequency of circadian oscillations," *Proc. Nat. Acad. Sci. USA*, 70 (1973), 2697–2701.

[264] Pittendrigh, C., J. Elliott, and T. Takamura, "The circadian component in photoperiodic induction," in *Photoperiodic Regulation of Insect and Molluscan Hormones*, eds. R. Porter and G. Collins (London: Pitman, 1984), pp. 26–47.

[265] Pittendrigh, C. and S. Daan, "Circadian oscillations in rodents: A systematic increase in their frequency with age," *Science*, 186 (1974), 548–50.

[266] Pittendrigh, C. and S. Daan, "A functional analysis of circadian pacemakers in nocturnal rodents. IV. Entrainment: Pacemaker as clock," *J. Comp. Physiol.*, 106 (1976), 291–331.

[267] Pittendrigh, C. and S. Daan, "A functional analysis of circadian pacemakers in nocturnal rodents. V. Pacemaker structure: A clock for all seasons," *J. Comp. Physiol.*, 106 (1976), 333–55.

[268] Pittendrigh, C. and S. Daan, "A functional analysis of circadian pacemakers in nocturnal rodents. I. The stability and lability of spontaneous frequency," *J. Comp. Physiol.*, 106 (1976), 223–52.

[269] Pittendrigh, C. and D. Minis, "Circadian systems: Longevity as a function of circadian resonance in *Drosophila melanogaster*," *Proceedings of the National Academy of Science USA*, 69 (1972), 137–39.

[270] Postlewaite, S. *Human Sexuality*. (Philadelphia, PA: W.B. Saunders Co., 1976), p. 14.

[271] Poulton, A., J. English, A. Symons, and J. Arendt, "Changes in plasma concentrations of LH, FSH and Prolactin in ewes receiving melatonin and short-photoperiod treatments to induce early onset of breeding activity," *J. Endocr.*, 112 (1987), 103–11.

[272] Pratt, B. and J. Takahashi, "Alpha-2 adrenergic regulation of melatonin release in chick pineal cell cultures," *Journal of Neuroscience*, 7 (1987), 3665–74.

[273] Ralph, C., S. Binkley, S. MacBride, and D. Klein, "Regulation of pineal rhythms in chickens: Effects of blinding, constant light, constant dark, and superior cervical ganglionectomy," *Endocrinology*, 97 (1975), 1373–78.

[274] Rawson, K. "Effects of tissue temperature on mammalian activity rhythms," *Cold Spring Harbor Symposia on Quantitative Biology*, 25 (1960), 105–13.

[275] Redman, J., S. Armstrong, and K. Ng, "Free-running activity rhythms in the rat: entrainment by melatonin," *Science*, 219 (1983), 1089–91.

[276] Reilly, K. and S. Binkley, "The menstrual rhythm," *Psychoneuroendocrinology*, 6 (1981), 181–84.

[277] Reinberg, A., Y. Motohashi, P. Bourdeleau, P. Andlauer, F. Levi, and A. Bicakova-Rocher, *Eur. J. Appl Physiol.*, 57 (1988), 15–25.

[278] Reiter, R., B. Richardson, L. Johnson, B. Ferguson, and D. Dinh, "Pineal melatonin rhythm: Reduction in aging Syrian hamsters," *Science*, 210 (1980), 1372–74.

[279] Rensing, L., W. Taylor, J. Dunlap, and J. Hastings, "The effects of protein synthesis inhibitors on the *Gonyaulax* clock. II. The effect of cycloheximide on ultrastructural parameters," *J. Comp. Physiol.*, 138 (1980), 9–18.

[280] Reppert, S. "Maternal coordination of fetal biological clock *in utero*," *Science*, 220 (1983), 969–71.

[281] Richter, C. "Biological clocks in medicine and psychiatry: shock-phase hypothesis," *Proc. Nat. Acad. Sci. USA*, 46 (1960), 1506–30.

[282] Richter, C. *Biological Clocks in Medicine and Psychiatry* (Springfield, IL: Charles C. Thomas, Publisher, 1965), 109 pages.

[283] Richter, C.P. "Sleep and activity: Their relation to the 24-hour clock," *Sleep and Altered States of Consciousness*, Assoc. for Research in Nervous and Mental Disorders, in *The Psychobiology of Curt Richter*, ed. E. Blass (Baltimore, MD: York Press, 1967), pp. 128–47.

[284] Richter, C.P. "Discovery of fire by man—Its effects on his 24-hour clock and intellectual and cultural evolution," *The Johns Hopkins Medical Journal*, 141 (1977), 47–61.

[285] Riebman, J. and S. Binkley, "Regulation of pineal glands of chickens: Organ culture," *Comp. Biochem. Physiol.*, 63C (1979), 93–98.

[286] Roberts, S. *Circadian Activity Rhythms in Cockroaches*, Ph.D. thesis, Princeton University, 1959.

[287] Roberts, S. "Circadian activity rhythms in cockroaches. I. The free-running rhythm in steady state," *J. Cell. and Comp. Physiol.*, 55 (1960), 99–110.

[288] Roberts, S. "Circadian activity rhythms in cockroaches. II. Entrainment and phase shifting," *J. Cell. and Comp. Physiol.*, 59 (1962), 175–86.

[289] Roberts, S. "Photoreception and entrainment of cockroach activity rhythms," *Science*, 148 (1965), 958–59.

[290] Roberts, S. "Circadian activity rhythms in cockroaches. III. The role of endocrine and neural factors," *J. Cell. Physiol.*, 67 (1966), 473–86.

[291] Roberts, S. "Circadian periodicity in cockroaches altered by high frequency light-dark cycles," *J. Comp. Physiol.*, 146 (1982), 255–59.

[292] Roberts, S. "Circadian rhythms in cockroaches. Effects of optic lobe lesions," *J. Comp. Physiol.*, 88 (1974), 21–30.

[293] Robertson, L. and J. Takahashi, "Circadian clock in cell culture I: Oscillation of melatonin release from dissociated chick pineal cells in flow-through microcarrier culture," *Journal of Neuroscience*, 8 (1988), 12–21.

[294] Robertson, L. and J. Takahashi, "Circadian clock in cell culture: II. In vitro photic entrainment of melatonin oscillation from dissociated chick pineal cells," *Journal of Neuroscience*, 8 (1988), 22–30.

[295] Roenneberg, T. and J. Hastings, "Two photoreceptors control the circadian clock of a unicellular alga," *Naturwissenschaften*, 75 (1988), 206–07.

[296] Rose, J. K. *The Body in Time* (New York: John Wiley & Sons, Inc., 1988), 237 pages.

[297] Rosenthal, N., F. Jacobsen, D. Sack, J. Arendt, S. James, B. Parry, and T. Wehr, "Atenolol in seasonal affective disorder: A test of the melatonin hypothesis," *Am. J. Psychiatry*, 145 (1986), 52–56.

[298] Rosenthal, N. and T. Wehr, "Seasonal affective disorders," *Psychiatric Annals*, 17 (1987), 670–74.

[299] Rowan, W. "Relation of light to bird migration and developmental changes," *Nature*, 115 (1925), 494–95.

[300] Sack, D., N. Rosenthal, B. Parry, and T. Wehr, "Biological rhythms in psychiatry," in *Psychopharmacology*, ed. H. Meltzer (New York: Raven Press, 1987), pp. 669–85.

[301] Sasaki, Y., N. Murakami, and K. Takahashi, "Critical period for the entrainment of the circadian rhythm in blinded pups by dams," *Physiology & Behavior*, 33 (1984), 105–09.

[302] Saunders, D. *An Introduction to Biological Rhythms* (Glasgow and London: Blackie, 1970), 170 pages.

[303] Saunders, D. *Insect Clocks* (Oxford: Pergamon Press, 1976), 279 pages.

[304] Schwartz, W. and H. Gainer, "Suprachiasmatic nucleus: Use of ^{14}C-labeled deoxyglucose uptake as a functional marker," *Science*, 197 (1977), 1089–91.

[305] Sherer, M., H. Weingartner, S. James, N. Rosenthal, "Effects of melatonin on performance testing in patients with seasonal affective disorder," *Neuroscience Letters*, 58 (1985), 277–82.

[306] Siffre, M. "Cave beyond time and transmeridian flight operation No. 1," Booklet published by Institute Francais de Speleologie, 1971.

[307] Siffre, M. "Six months alone in a cave," *Natl. Geogr. Mag.*, 147 (1975), 426–35.

[308] Smith, R. and R. Konopka, "Circadian clock phenotypes of chromosome aberrations with a breakpoint at the per locus," *Mol. Gen. Genet.*, 183 (1981), 243–51.

[309] Sonis, W., A. Yellin, B. Garfinkel, and H. Hoberman, "The antidepressant effect of light in seasonal affective disorder of childhood and adolescence," *Psychopharmacology Bulletin*, 23 (1987), 360–63.

[310] Steinlechner, S., A. Buchberger, and G. Heldmaier, "Circadian rhythms of pineal N-acetyltransferase activity in the Djungarian hamster", *Phodopus sungorus*, in response to seasonal changes of natural photoperiod," *J. Comp. Physiol.*, A. 160 (1988), 593–97.

[311] Stephan, F. and I. Zucker, "Circadian rhythms in drinking behavior and locomotor activity of rats are eliminated by hypothalamic lesions," *Proc. Nat. Acad. Sci. USA*, 69 (1972), 1583–86.

[312] Stetson, M. and D. Tay, "Time course of sensitivity of golden hamsters to melatonin injections throughout the day," *Biology of Reproduction*, 29 (1983), 432–38.

[313] Stetson, M. and M. Watson-Whitmyre, "Nucleus suprachiasmaticus: The biological clock in the hamster?," *Science*, 191 (1976), 197–99.

[314] Stetson, M. and M. Watson-Whitmyre, "Physiology of the pineal and its hormone melatonin in annual reproduction in rodents," in *The Pineal Gland*, ed. R. Reiter (New York: Raven Press, 1984), pp. 109–53.

[315] Strogatz, S. *The Mathematical Structure of the Human Sleep-Wake Cycle*. (New York: Springer-Verlag, New York, Inc., 1986), 230 pages.

[316] Strumwasser, F. "The demonstration and manipulation of a circadian rhythm in a single neuron," in *Circadian Clocks*, ed. J. Aschoff, (Amsterdam: North-Holland Pub. Co., 1965), pp. 442–62.

[317] Strumwasser, F. "The cellular basis of behavior in *Aplysia*," *J. Psychiat. Res.*, 8 (1971), 237–57.

[318] Sulzman, F., V. Gooch, K. Homma, an J. Hastings, "Cellular autonomy of the *Gonyaulax* circadian clock," *Cell Biophysics*, 4 (1982), 97–103.

[318b] Sulzman, F., D. Ellman, C. Fuller, M. Moore-Ede, and G. Wassmer, "*Neurospora* circadian rhythms in space: A reexamination of the endogenous-exogenous question." *Science*, 225 (1984), 232–234.

[319] Sweeney, B. "The photosynthetic rhythm in single cells of *Gonyaulax polyedra*," *Cold Spring Harbor Symposia on Quantitative Biology*, 25 (1960), 145–58.

[320] Sweeney, B. *Rhythmic Phenomena in Plants* (New York: Academic Press, 1969), 147 pages.

[321] Sweeney, B. "The potassium content of *Gonyaulax polyedra* and phase changes in the circadian rhythm of stimulated bioluminescence by short exposures to ethanol and valinomycin," *Plant Physiol.*, 53 (1974), 337–42.

[322] Sweeney, B. "Circadian time-keeping in eukaryotic cells, models and hypotheses," *Progress in Phycological Research*, 2 (1983), 189–225.

[323] Sweeney, B. "Freeze-fracture studies of the thecal membranes of *Gonyaulax polyedra*: Circadian changes in the particles of one membrane face," *The Journal of Cell Biology*, 68 (1976), 451–61.

[324] Sweeney, B. "Bright light does not immediately stop the circadian clock of *Gonyaulax*," *Plant Physiol.*, 64 (1979), 341–44.

[325] Sweeney, B. and J. Hastings, "Characteristics of the diurnal rhythm of luminescence in *Gonyaulax polyedra*," *J. Cell. Comp. Physiol.*, 49 (1957), 115–28.

[326] Sweeney, B. M. and J. Hastings, "Effects of temperature upon diurnal rhythms," *Cold Spring Harbor Symposia on Quantitative Biology*, 25 (1960), 87–113.

[327] Takahashi, J. *Neural and endocrine regulation of avian circadian systems*, Ph.D. diss, University of Oregon, (1981), 187 pages.

[328] Takahashi, J. and F. Turek, "Anisomycin, an inhibitor of protein synthesis,

perturbs the phase of a mammalian circadian pacemaker," *Brain Research*, 405 (1987), 199–203.

[329] Takahashi, J. and M. Menaker, "Entrainment of the circadian system of the house sparrow: A population of oscillators in pinealectomized birds," *J. Comp. Physiol.*, 146 (1982), 255–59.

[330] Takahashi, J. and M. Menaker, "Role of the suprachiasmatic nuclei in the circadian system of the house sparrow, *Passer domesticus*," *The Journal of Neurosciences*, 2 (1982), 815–28.

[331] Takahashi, J., H. Hamm, and M. Menaker, "Circadian rhythms of melatonin release from individual superfused chicken pineal glands *in vitro*," *Proc. Natl. Acad. Sci. USA*, 77 (1980), 2319–22.

[332] Takahashi, K., C. Hayafuji, and N. Murakami, "Foster mother rat entrains circadian adrenocortical rhythm in blinded pups," *Am. J. Physiol*, 243 (1982), E443–49.

[333] Takahashi, K. and T. Deguchi, "Entrainment of the circadian rhythms of blinded infant rats by nursing mothers," *Physiology & Behavior*, 31 (1983), 373–78.

[334] Takahashi, K., N. Murakami, C. Hayafuji, and Y. Sasaki, "Further evidence that circadian rhythm of blinded rat pups is entrained by the nursing dam," *Am. J. Physiol.*, 246 (1984), R359–63.

[335] Takeda, M., Y. Endo, H. Saito, M. Nishimura, and J. Nishiitsutsuji-Uwo, "Neuropeptide and monoamine immunoreactivity of the circadian pacemaker in *Periplaneta*," *Biomedical Research*, 6 (1985), 395–406.

[336] Tamarkin, L., W. Westrom, A. Hamill, and B. Goldman, "Effect of melatonin on the reproductive systems of male and female Syrian hamsters: A diurnal rhythm in sensitivity to melatonin," *Endocrinology*, 99 (1976), 1534–41.

[337] Taylor, W. and J. Hastings, "Minute-long pulses of anisomycin phase-shift the biological clock in *Gonyaulax* by hours," *Naturwissenschaften*, 69 (1982), 94–95.

[338] Taylor, W, V. Gooch, and J. Hastings, "Period shortening and phase shifting effects of ethanol on the *Gonyaulax* glow rhythm," *J. Comp. Physiol.*, 130 (1979), 355–58.

[339] Thomas, R. *The Old Farmer's Almanac* (Dublin, NH: Yankee Publishing, Inc., 1988), 231 pages.

[340] Thoreau, H. *Walden* (Boston MA: Ticknor and Shields, 1854), 357 pages.

[341] Tokura, H. and J. Aschoff, "Circadian rhythms of locomotor activity in the squirrel monkey, *Saimiri sciureus*, under conditions of self-controlled light-dark cycles," *Japanese Journal of Physiology*, 29 (1979), 151–57.

[342] Treolar, A., R. Boynton, B. Behn, and B. Brown, "Variation of the human menstrual cycle through reproductive life," *Int. J. Fert.*, 12 (1967), 77–126.

[343] Truman, J. "Physiology of insect rhythms. II. The silk moth brain as the location of the biological clock controlling eclosion," *J. Comp. Phys.*, 81 (1972), 99–114.

[344] Turek, F. "Circadian principles and design of rotating shift work schedules," *Am J. Physiol.*, 251 (1986), R636–39.

[345] Turek, F. and S. Losee-Olson, "A benzodiazepine used in the treatment of insomnia phase-shifts the mammalian circadian clock," *Nature*, 321 (1986), 167–68.

[346] Turek, F. and S. H. Losee-Olson, "Dose response curve for the phase-shifting effect of triazolam on the mammalian circadian clock," *Life Sciences*, 40 (1987), 1033–38.

[347] Turek, F., J. McMillan, and M. Menaker, "Melatonin: Effects on the circadian locomotor rhythm of sparrows," *Science*, 194 (1976), 1441–43.

[348] Underwood, H. "Circadian organization in the lizard *Anolis carolinensis*: A multioscillator system," *J. Comp. Physiol.*, 152 (1983), 265–74.

[349] Underwood, H. "Circadian organization in Japanese quail," *J. Exp. Zoo.*, 232 (1984), 557–66.

[350] Underwood, H. "Circadian rhythms in lizards: Phase response curve for melatonin," *Journal of Pineal Research*, 3 (1986), 187–96.

[351] Underwood, H. and Gerard Groos, "Vertebrate circadian rhythms: Retinal and extraretinal photoreception," *Experientia*, 38 (1982), 1013–21.

[352] Underwood, H. and M. Harless, "Entrainment of the circadian activity rhythm of a lizard to melatonin injections," *Physiol. Behav.*, 35 (1985), 267–70.

[353] Underwood, H. and M. Menaker, "Photoperiodically significant photoreception in sparrows: Is the retina involved?," *Science*, 167 (1970), 298–301.

[354] Vaněček, J., A. Pavlik, and Helena Illnerová, "Hypothalamic melatonin receptor sites revealed by autoradiography," *Brain Research*, 435 (1987), 359–62.

[355] Vanden Dreissche, T. "Structural and functional rhythms in the chloroplasts of *Acetabularia*: Molecular aspects of the circadian system," in *Biochronometry*, ed. M. Menaker (Washington D.C.: National Academy of Sciences, 1971), pp. 612–22.

[356] Van Vleck, J. H. and D. Middleton, "The spectrum of clipped noise," *Proceedings of the IEEE*, 54 (1966), 2–19.

[357] Vaughan, G. "Melatonin in humans," in *Pineal Research Reviews*, ed. R. Reiter (New York: Alan R. Liss, Inc, 1984), pp. 141–201.

[358] Vaughan, G. and R. Reiter, "The Syrian hamster pineal gland responds to isoproterenol *in vivo* at night," *Endocrinology*, 120 (1987), 1682–84.

[359] Vaughan, G., B. Pruitt, and A. Mason, "Nyctohemeral rhythm in melatonin response to isoproterenol *in vitro*: Comparison of rats and Syrian hamsters," *Comp. Biochem. Physiol.*, 87C (1987), 71–74.

[360] Vernikos-Danellis, J., C. Winget, and J. Beljan, "The effect of antiemetic medication on human circadian rhythms," in *Chronopharmacology and Chronotherapeutics*, eds. C. Walker, C. Winget, and K. Soliman (Tallahassee, FL: Florida A & M University, 1981), pp. 401–11.

[361] Vivien-Roels, B., J. Arendt, and J. Bradke, "Circadian and circannual fluctuations of pineal indoleamines (serotonin and melatonin) in *Testudo hermanni* Gmelin (Reptilia, Chelonia)," *Gen. Comp. Endocrinol.*, 37 (1979), 179–210.

[362] Vivien-Roels, B. and A. Meinel, "Seasonal variation of serotonin in the pineal complex and the lateral eye of *Lampetra planeri* (Cyclostoma, Petromyzontidae)," *Gen. Comp. Endocrinol.*, 50 (1983), 313–23.

[363] Volknandt, W. and R. Hardeland, "Circadian rhythmicity of protein synthesis in the dinoflagellate, *Gonyaulax polyedra*: A biochemical and radioautographic investigation," *Comp. Biochem. Physiol.*, 77B (1984), 493–500.

[364] Wainwright, S. "Diurnal cycles in serotonin acetyltransferase activity and cyclic GMP content of cultured chick pineal cells," *Nature*, 285 (1980), 478–80.

[365] Wainwright, S. and L. Wainwright, "The relationship between variations in levels of serotonin acetyltransferase activity and cGMP content in cultured chick pineal glands," *Can. J. Biochemistry*, 59 (1981), 593–601.

[366] Walker, C., C. Winget, and K. Soliman, eds., *Chronopharmacology and Chronotherapeutics* (Tallahassee, FL: Florida A & M University Press, 1981), 417 pages.

[367] Walz, B., A. Walz, and B. Sweeney, "A circadian rhythm in RNA in the dinoflagellate *Gonyaulax polyedra*," *J. Comp. Physiol.*, 151 (1983), 207–13.

[368] Ward, R. *The Living Clocks* (New York: Mentor Books, 1971), 351 pages.

[369] Webb, W. "Twenty-four hour sleep cycling," in *Sleep*, ed. A. Kales (Philadelphia, PA: J. B. Lippincott Co., 1969), pp. 53–65.

[370] Weber, A. and N. Adler, "Delay of constant light-induced persistent vaginal estrus by 24-hour time cues in rats," *Science*, 204 (1979), 323–25.

[371] *Webster's New Universal Unabridged Dictionary* (New York: Simon and Schuster, 1979).

[372] Wehr, T., F. Jacobsen, D. Sack, J. Arendt, L. Tamarkin, and N. Rosenthal, "Phototherapy of seasonal affective disorder," *Arch. Gen. Psychiatry*, 43 (1986), 870–75.

[373] Wehr, T., D. Sack, N. Rosenthal, W. Duncan, and J. Gillin, "Circadian rhythm disturbances in manic-depressive illness," *Federation Proceedings*, 42 (1983), 2809–14.

[374] Wehr, T., A. Wirz-Justice, F. Goodwin, W. Duncan, and J. Gillin, "Phase advance of the circadian sleep-wake cycle as an antidepressant," *Science*, 206 (1979), 710–13.

[375] Wells, H. *The War of the Worlds*, (New York, Signet Classics, 1898), p. 142.

[376] Wever, R. "A mathematical model for circadian rhythms," in *Circadian Clocks*, ed. J. Aschoff (Amsterdam: North-Holland Pub. Co., 1965), pp. 47–63.

[377] Wever, R. "Pendulum versus relaxation oscillation," in *Circadian Clocks*, ed. J. Aschoff (Amsterdam: North-Holland Pub. Co., 1965), pp. 74–83.

[378] Wever, R. "Fractional desynchronization of human circadian rhythms: A method for evaluating entrainment limits and functional interdependencies," *Pflügers Arch.*, 396 (1983), 128–37.

[379] Wever, R. "Sex differences in human circadian rhythms: Intrinsic periods and sleep fractions," *Experientia*, 40 (1984), 1226–34.

[380] Wever, R. "Toward a mathematical model of circadian rhythmicity," in *Mathematical Models of the Circadian Sleep-Wake Cycle*, eds. M. Moore-Ede and C. Czeisler (New York: Raven Press, 1984), pp. 17–79.

[381] Wever, R. "Use of light to treat jet lag: Differential effects of normal and bright artificial light on human circadian rhythms," *Annals of the New York Academy of Science*, 453 (1985), 283–304.

[382] Wever, R., J. Polasek, and C. Wildgruber, "Bright light affects human circadian rhythms," *Pflügers. Arch.*, 396 (1983), 85–87.

[383] Winfree, A. T. "Corkscrews and singularities in fruitflies: resetting behavior of the circadian eclosion rhythm," in *Biochronometry*, ed. M. Menaker (Washington, D.C.: National Academy of Sciences, 1971), pp. 81–109.

[384] Winfree, A. T. *The Geometry of Biological Time*, New York: Springer-Verlag New York, Inc., 1980), 530 pages.

[385] Winfree, A. T. "Benzodiazepines set the clock," *Nature*, 321 (1986), 114–15.

[386] Winfree, A. T. *The Timing of Biological Clocks* (New York: Scientific American Books, 1987) 200 pages.

[387] Wright, J., M. Aldhous, C. Franey, J. English, and J. Arendt, "The effects of exogenous melatonin on endocrine functions in man," *Clinical Endocrinology*, 24 (1986), 375–82.

[388] Zatz, M., D. Mullen, and J. Moskal, "Photoendocrine transduction in cultured chick pineal cells: Effects of light, dark, and potassium on the melatonin rhythm," *Brain Research*, 438 (1988), 199–215.

[389] Zimmerman, N. and M. Menaker, "The pineal gland: A pacemaker within the circadian system of the house sparrow," *Proc. Natl. Acad. Sci. USA*, 76 (1979), 999–1003.

[390] Zimmerman, W. and T. Goldsmith, "Photosensitivity of the circadian rhythm and of visual receptors in carotenoid-depleted *Drosophila*, *Science*, 171 (1971), 1167–69.

[391] Zimmerman, W. and D. Ives, "Some photophysiological aspects of circadian rhythmicity in *Drosophila*," In *Biochronometry*, ed. M. Menaker (Washington, D.C.: National Academy of Sciences, 1971), pp. 381–91.

[392] Zimmerman, W., C. Pittendrigh, and T. Pavlidis, "Temperature compensation of the circadian oscillation in *Drosophila pseudoobscura* and its entrainment by temperature cycles," *J. Insect Physiol.*, 14 (1968), 669–84.

[393] Zucker, I. "Pineal gland influences period of circannual rhythms of ground squirrels," *Am. J. Physiol.*, 249 (1985), R111–15.

[394] Zucker, I., M. Bosches, and J. Dark, "Suprachiasmatic nuclei influence circannual and circadian rhythms of ground squirrels," *Am. J. Physiol.*, 13 (1983), R472–80.

GLOSSARY*

a.m. am, A.M., AM, before noon, used to designate the time from midnight to noon

acrophase phase angle of the maximal value of a sine function fitted to the raw data of a rhythm; a measure of the time of the peak of an oscillation

ad libitum used in this book to denote free choice of feeding or light in experiments

advancing phase shift phase shift where one or more period is shortened; conventionally given a positive (+) sign

aftereffects characteristics of a rhythm which derive from pretreatment conditions, such as transients in phase or changes in period length, e.g., knees in sparrows' rhythms

alpha (α) activity time, the duration of activity

amplitude (of a rhythm) maximum minus minimum value, excursion

antigonadal inhibitory to the reproductive system, causes the testes and/or ovaries to decrease in size or weight or to become atrophic

---

* Glossaries from books in the list of references were the source of these definitions [9, 14, 104, 239, 250, 302, 303].

arrhythmic   lacking a discernible rhythm; sometimes results from pinealec-
     tomy or constant light

Aschoff's rule   nocturnal animals have DD periods less than LL periods,
     diurnal animals have DD periods greater than LL periods; period of the
     freerunning oscillation shortens as light intensity increases in day-active
     animals and lengthens as light intensity increases in night-active
     animals

asynchrony   lack of synchronization between two rhythms

beta receptor   theoretical membrane receptor for catecholamines (e.g., nor-
     epinephrine) that is defined pharmacologically (based on responses to
     stimulating and blocking agents)

biological clock   mechanism within an organism that is capable of generating
     repeated cycles (oscillations, rhythms), whose period is relatively insen-
     sitive to temperature, and which can be synchronized by environmental
     stimuli

cAMP   cyclic adenosine 3', 5' monophosphate, cyclic AMP

circadian rhythm   a rhythm with a period of about 24 hours

circadian rule   activity time increases with light intensity increase in diurnal
     animals; activity time decreases with light intensity increase in noctur-
     nal animals

circannual rhythm   also circannian rhythm; an endogenous rhythm with a
     period of about but not exactly a year

clock   "an instrument for the measurement of time by the motion of its parts,
     indicating the time by the hours, minutes, and often seconds, by hands
     which move upon a dial plate. It usually consists of a frame containing a
     train of toothed wheels operated by springs or weights and regulated by a
     pendulum or balance wheel. It differs from a watch in that it is not worn
     or carried about in the pocket." [371]

crepuscular   active at twilight or just before dawn

critical photoperiod   the duration of light in excess of which a photoperiodic
     response is that of a "long" day

c.t.   also Ct, ct, CT; circadian time, time from the organism's point of view in
     which the organism's circadian cycle is set "equal" to 24 and each 1/24 of
     the cycle is considered to be an "hour." For diurnal animals, c.t. 0 by
     convention = onset of activity, and for nocturnal animals, c.t. 12 by
     convention = onset of activity.

cycle   a period of time during which a sequence of events occurs—one such
     period is a cycle but the use of the word implies that the sequence of
     events repeats

D   dark-time

DD   constant dark

dark-time   night in natural lighting; the time during which lights are turned off in experiments

day   can mean either the period of time between dawn and dusk, or it can mean 24 hours

delaying phase shift   phase shift resulting from lengthening of one or a few periods; conventionally designated with a minus ($-$) sign

desynchronization   "loss of synchronization between two or more rhythms so that they show independent periods" [239]

diapause   period of delayed growth or development, inactivity, and reduced metabolism

diurnal   active in the day-time

endocrine gland   a ductless gland which secretes chemicals (hormones) into the blood

entrainment   synchronization of a rhythm by a repetitive signal (e.g. the recurring light-dark cycle)

freerun   repetitive cycles of a rhythm in the absence of a synchronizing signal; the period of the *freerun*, tau, is considered to be the innate period of the rhythm (= free run)

frequency   reciprocal of period; to determine the frequency, divide one by the period length (e.g., if period = 24.0h, then frequency = $1/24$ = 0.04166 cycles/day)

g   gram

h   hour, hr., hr, 60 minutes

5-HT   5-hydroxytryptamine; serotonin; precursor of melatonin

hormone   chemical product of a ductless gland carried to its *target* via the blood; less rigorously used to denote any *chemical messenger*

interval timer   a nonrepetitive timer such as an hourglass

in vitro   outside the animal, as in organ culture, cell culture, or superfusion culture

in vivo   in the living animal

jet lag   the syndrome of symptoms that occur when a person's physiological rhythms are out of phase with the environmental light-dark cycle and each other (e.g., after transmeridian flights) [250]

L   light-time

LD14 : 10   light-dark cycle consisting of 14 hours of light in alternation with 10 hours of dark; often designated 14L : 10D

LDLD1 : 6 : 1 : 16   a skeleton light-dark cycle consisting of two 1h pulses of light alternating with 6 and 16h of dark; the cycle is the skeleton of LD8 : 16 or of LD18 : 6

LL   constant light

m   meter

M   midnight; or, in context, M phase of cell division

masking   effect on the overt circadian rhythm in a positive (enhancing) or negative (depressing) manner not related to the process of entrainment due to periodic changes in the intensity of illumination. For example, this might be additional activity in response to light in a sparrow or increased activity in a hamster during a dark pulse [20].

mean value   (of an oscillation), average of all values of the parameter measured with one period

melanophore   pigment-containing cell in the skin

melatonin   M, MT, MEL; methoxyindole hormone of the pineal gland

min   minute, min., 60 seconds, $1/60$ hour

N   used in context, N = noon when time is referred to, N also is used to indicate number of samples in experiments

NAT   N-acetyltransferase activity; NAT is a pineal enzyme which converts serotonin to N-acetylserotonin; distinguished by its marked circadian rhythm and sensitivity to light and dark

NE   norepinephrine, a catecholamine that stimulates melatonin synthesis in some species

night   the darkness that occurs between dusk and dawn

nocturnal   active in the night-time

nychthemeron   an entire day, 24 hours

oscillation   a cycle

pacemaker   driving oscillation, a rhythm that controls other rhythms

period (τ)   the length of a cycle; designated by the greek letter tau; "time after which a definite phase of the oscillation reoccurs" [9]

phase (ϕ)   (of a rhythm) the "instantaneous state of an oscillation with a period, represented by the value of the variable and all its time derivatives" [9]; phase specifies the relationship between an event and something else (e.g., time according to a clock) and requires specification of phase reference points

phase angle the difference in time between the phase of an event and the phase of another event (e.g., the difference in time between an organism's activity onset and the time of lights-on); requires specification of the event's reference points; can be given in units of time if the period is specified or in fractions of a period

phase reference points specification of events, required for determining phase and phase angle

phase response curve plot of phase shifts in response to pulses plotted versus time the pulse was given

phase shift ($\Delta\phi$) "a single displacement of an oscillation along the time-axis" [9]

photoperiod light-time, length of the light-time; daylength

photoperiodic phenomenon that is responsive to daylength (or nightlength)

photoreceptor a structure that detects light

p.m. pm, P.M., PM, afternoon; the time from noon to midnight

PRC phase response curve

Q_{10} the temperature coefficient = rate at temperature (X + 10°C) / (rate at temperature X) [250]

range of entrainment range of Zeitgeber periods with which a rhythm can synchronize

rapid plummet drop in N-acetyltransferase or melatonin that occurs in response to light at night with a halving time of five minutes or less

recrudescence (e.g., of testes) growth, increase in size, renewal of function; normal part of a seasonal cycle

refractory period portion of time including the late subjective night and early subjective day when melatonin synthesis (N-acetyltransferase) cannot be stimulated by placing the organism in dark

regression (e.g., of testes) involution, reduction in size, diminution of function; normal part of a seasonal cycle

resonance experiment a protocol in which groups are subjected to cycles whose periods are multiples (24h, 48h, 72h) and nonmultiples of 24h (36h, 60h)

retinohypothalamic tract RHT; nerve tract from the eye to the hypothalamus

rho (ρ) rest-time, the duration of rest

rhythm a pattern of events that recurs; a series of cycles

RIA radioimmunoassay; a method of measuring a hormone or other molecule that involves binding characteristics of specific antibodies

scotoperiod nightlength; length of the dark-time

SCG superior cervical ganglion

SCN suprachiasmatic nucleus, nucleus suprachiasmaticus, a region of the hypothalamus, bilaterally paired

sidereal time time expressed with reference to the stars; the sidereal day is the time elapsed between two successive passages of the vernal equinox over the upper meridian; one rotation of the earth; 23h, 56 min, and 11.5 sec of mean solar time

Silastic capsules tubes with plugged ends filled with a substance (e.g., melatonin crystals); a means of administering a hormone continuously

singularity (T*S*) a light pulse of specified intensity and duration (S*) imposed at a time (T*) after the LL/DD transition, which puts the clock in a nonoscillatory state (i.e., an annihilating pulse that stops the clock). For example, for a population of Drosophila pseudoobscura, T*S* for the eclosion rhythm is a 50-second pulse of dim blue light (10 microWatts/cm^2) 6.8h after the LL/DD transition (i.e., at about circadian time 18–19) [303, 383, 386].

skeleton photoperiod a light-dark cycle in which dawn and dusk are represented by two separate light periods, e.g, LDLD1 : 7 : 1 : 15 (the skeleton of LD9 : 15 or of LD17 : 7); if the two light periods are of equal length, then the skeleton is considered to be symmetric

solar time time measured with reference to the earth's motion in relation to the sun

subjective day the time during which the organism exhibits the behavior usually associated with light in an LD cycle; for example, a diurnal sparrow is usually active in its subjective day while a nocturnal hamster usually rests in its subjective day

subjective night the time during which the organism exhibits the behavior usually associated with dark in an LD cycle; for example, a sparrow usually rests in its subjective night while a hamster is active in its subjective night

subsensitivity reduced ability of melatonin synthesis to respond to stimulation, e.g., by isoproterenol; occurs after prolonged stimulation

superior cervical ganglia (SCG) (ganglion) a pair of sympathetic ganglia located in the neck; relay neural signals to the pineal gland (SCG)

supersensitivity increased ability of melatonin synthesis to respond to stimulation, e.g., by isoproterenol; occurs after stimulation deprivation, e.g. by denervation

suprachiasmatic nuclei (nucleus) region of the hypothalamus thought to generate circadian rhythm information that is conveyed neurally to the pineal gland in some species (e.g., rats); bilateral

T period of a Zeitgeber cycle

target cells, tissues, or organs upon which a hormone acts

tau (τ) period of a circadian rhythm

T experiments experiments in which the period of the Zeitgeber cycle is varied experimentally so that it is not 24h but is some other amount, e.g., 23h or 25h

transducer mechanism for converting one type of signal into another type of signal (e.g., nerve signal into endocrine signal)

transients "temporary oscillatory states between two steady states" [9]; e.g., the cycles observed when resetting occurs in response to shifting the phase of the entraining cycle or in response to light, dark, or temperature pulses

Zeitgeber synchronizer, entraining agent, time giver, time signal, time cue, e.g., dawn or dusk

ANNOTATED BIBLIOGRAPHY

To assist in locating further reading, I have compiled the following list of some of the books published about circadian rhythms and closely related topics. The earliest book on this list is that of Bünning (1958). Unfortunately, some of these books are now out of print.

ASCHOFF, J., ed. *Circadian Clocks*. Amsterdam: North-Holland Pub. Co., 1965. 44 articles make up the proceedings of the Feldafing summer school in 1964, Glossary in English and German. 479 pages.

ASCHOFF, J., ed. *Biological Rhythms*. vol. 4 of *Handbook of Behavioral Biology*. 27 invited chapters and Glossary. New York: Plenum Press, 1981. 562 pages.

ASCHOFF, J., S. DAAN, and G. A. GROOS. *Vertebrate Circadian Systems: Structure and Physiology*. Contributions from a meeting in Schloss Ringberg in 1980. 37 articles, 154 figures. New York: Springer-Verlag New York, Inc., 1982. 363 pages.

AYENSU, E. S., and P. WHITFIELD. *The Rhythms of Life*. New York: Crown Publishers, Inc., 1981. 199 pages. A volume with pleasing illustrations written for the general public; includes biological clocks and calendars but also other rhythms exhibited by living things as in motion and music.

BENNETT, M. F. *Living Clocks in the Animal World*. Springfield, IL: Charles C. Thomas, Publisher, 1974. 221 pages. A monograph reviewing biological clocks which focuses on crabs, honey bees, earthworms, mollusks, amphibians.

Biological Clocks. vol. XXV of the Cold Spring Harbor Symposia on Quantitative Biology. Baltimore, MD: Waverly Press. 51 symposium papers. Terminology on p. 160. 524 pages.

BRADY, J. 1979. *Biological Clocks*. London: Edward Arnold Publishers Ltd., 1979. Seven chapters cover daily, tidal and annual rhythms; the exogenous or endogenous issue; circadian rhythms; celestial navigation and continuously consulted clocks; photoperiodism; and clock mechanisms. 60 pages.

BRADY, J., ed. *Biological Timekeeping*. New York: Cambridge University Press, 1982. Intended as a student text by authors from Seminar organized by the Society for Experimental Biology at Imperial College in London in 1980. A glossary and eleven invited chapters. 197 pages.

BROWN, F., J. W. HASTINGS, and J. D. PALMER. *The Biological Clock, Two Views*. New York: Academic Press, 1970. Discussion of contrasting exogenous and endogenous hypotheses for explaining rhythmic phenomena by their respective proponents. 94 pages.

BÜNNING, E. *Die Physiologische Urh*. Berlin: Springer Verlag, 1958. The first monograph in the field of circadian rhythms.

BÜNNING, E. *The Physiological Clock* (3rd ed.) New York: Springer-Verlag, 1973. An enduring monograph in English originally published in 1958 under the title *Die Physiologishche Uhr*. 167 pages.

CARPENTER, D. O. ed. *Cellular Pacemakers*. New York: John Wiley & Sons, Inc., 1982. Vol. 2 of Function in Normal and Disease States. 15 invited articles on cellular pacemakers in the nervous system, circadian rhythms, and what happens when pacemaking systems break down in disease and aging.

COLEMAN, R. M. *Wide Awake at 3:00 A.M. By Choice or By Chance?* New York: W. H. Freeman and Co. Publishers, 1986. Emphasis on human and sleep related aspects of circadian rhythms; Glossary. 195 pages.

CONROY, R., and J. MILLS. *Human Circadian Rhythms*. London: J. & A. Churchill, 1970. Chapters on methods, abnormal time schedules, and rhythms (endocrine, temperature, kidney, cardiovascular, respiratory, wakefulness, birth). Opens with a section discussing definitions. 945 references. 236 pages.

DECOURSEY, P. J., ed. *Biological Rhythms in the Marine Environment*. Columbia, SC: University of South Carolina Press, 1976. 20 research articles from a symposium at the Hobcaw House on the Belle W. Baruch Coastal Research Station near Georgetown. 283 pages.

EDMUNDS, L. N. *Cell Cycle Clocks*. New York: Marcel Dekker, Inc., 1984. 27 articles detailing cell division cycles and their temporal organization. 616 pages.

EDMUNDS, L. N. *Cellular and Molecular Bases of Biological Clocks: Models and Mechanisms for Circadian Timekeeping*. New York: Springer-Verlag New York, Inc., 1988. A monograph considering the state of the art of the molecular, cellular, and biochemical basis of circadian rhythmicity. The book is organized into six chapters (introduction, eukaryotic microorganisms, cell cycle clocks, experimental approaches, biochemical and molecular models, and general considerations). Useful study aids include 156 illustrations, a list of abbreviations, 57 pages of references, an author index, and a subject index. 497 pages.

EHRET, C. F., and L. W. SCANLON. *Overcoming Jet Lag*. New York: Berkley Pub. Group, 1983. Prescriptions for travelers hoping to reduce the time shifting effects of east-west travel. 160 pages.

ENRIGHT, J. T. *The timing of sleep and wakefulness: On the substructure and dynamics of the circadian pacemakers underlying the wake-sleep cycle.* A monograph with computer simulations. New York: Springer-Verlag New York, Inc., 1980. 263 pages.

FOLLETT, B. K., and D. E. FOLLETT. *Biological Clocks in Seasonal Reproductive Cycles.* New York: John Wiley & Sons, 1981. Proceedings of the 32nd Symposium of the Colston Research Society held in the University of Bristol in 1980; with a species index. 292 pages.

GLASS, L. *From Clocks to Chaos: The Rhythms of Life.* Princeton: Princeton University Press, 1988.

GWINNER, E. *Circannual Rhythms: Endogenous Annual Clocks in the Organization of Seasonal Processes.* New York: Springer-Verlag New York, Inc., 1986. An inventory of work on the subject with a systematic index and 73 figures of scientific data and models. Gwinner's book is volume 18 in the series of Zoophysiology. 154 pages.

HALBERG, F., ed. *Proceedings of the XII International Conference of the International Society for Chronobiology.* Milano, Italia: Il Ponte, 1975. Nearly 100 research articles from a 1975 conference held in Washington, DC. 782 pages.

HALBERG, F. *Glossary of Chronobiology.* Milano, Italia: Il Ponte, 1977. English/Italian. 188 pages.

HARKER, J. *The Physiology of Diurnal Rhythms.* New York: Cambridge University Press, 1964. A monograph covering the role of the environment, freerunning rhythms, phase shifting, and rhythm abnormalities. 114 pages.

HIROSHIGE, T., and K. HONMA, eds. *Circadian Clocks and Zeitgebers.* Sapporo, Japan: Hokkaido University Press, 1985. 17 research articles, Proceedings of the first Sapporo Symposium on Biological Rhythm, August 29–31, 1984. 191 pages.

KRIEGER, D. T. *Endocrine Rhythms.* New York: Raven Press, 1979. 12 invited chapters considering the "basic classification, causation, and properties of endocrine rhythms, and the neuroan· ·omy of the pathways involved in their regulation." 332 pages.

LUCE, G. G. *Biological Rhythms in Psychiatry and Medicine.* U.S. Government Printing Office, 1970. A National Institute of Mental Health report containing "compelling evidence that man is constructed not only of matter, but that he is temporally organized . . . with . . . significant implications for . . . mental and physical health." Public Health Service Publication No. 2088. 183 pages.

LUCE, G. G. *Body Time: Physiological rhythms and social stress.* New York: Bantam Books, Inc., 1971. A book for the general audience derived from the NIMH report. 441 pages.

MENAKER, M. ed. *Biochronometry.* Washington, DC: National Academy of Sciences, 1971. Proceedings of a symposium at Friday Harbor, Washington, in 1969. 40 research articles "that analyze circadian systems at the physiological level and that lead one to expect further rapid progress." 662 pages.

MEYERSBACH, H. VON, L. E. SCHEVING, AND J. E. PAULY, eds. *Biological Rhythms in Structure and Function.* New York: Alan R. Liss, 1981. 13 papers from the International Congress of Anatomy held in Mexico City in 1980. 241 pages.

MILLS, J. N. ed. *Biological Aspects of Circadian Rhythms.* London: Plenum Press, 1973. Glossary and eight reviews by nine authors. 319 pages.

MOORE-EDE, M. C., F. M. SULZMAN, and C. A. FULLER. *The Clocks that Time Us: Physiology of the Circadian Timing System.* Cambridge, MA: Harvard University Press, 1982. Introductory book with Glossary. 448 pages.

PALMER, J. D. *Biological Clocks in Marine Organisms:* The control of physiological and behavioral tidal rhythms ("bimodal oscillations with a period about 24.8h"). New York: John Wiley & Sons, Inc., 1974. Glossary. 173 pages.

PALMER, J., F. BROWN, and L. EDMUNDS. *An Introduction to Biological Rhythms.* Intended as a textbook for students and interested biologists, with a glossary. New York: Academic Press, 1976. 375 pages.

PAULY, J. E., L. E. SCHEVING. *Progress in Clinical and Biological Research V.* 227A. Advances in Chronobiology, Part A., Proceedings of the XVIIth International Conference of the International Society for Chronobiology. New York: Alan R. Liss, 1987. 528 pages.

PAULY, J. E., L. E. SCHEVING. *Progress in Clinical and Biological Research V.* 227B. Advances in Chronobiology, Part B., Proceedings of the XVIIth International Conference of the International Society for Chronobiology. New York: Alan R. Liss, 1987. 613 pages.

PAVLIDIS, T. *Biological Oscillators: Their Mathematical Analysis.* New York: Academic Press, Inc., 1973. Monograph on "biological oscillators and the mathematical techniques necessary for their investigation . . . accessible to anyone familiar with the basic concepts and the formalism of elementary differential equations and linear algebra." 207 pages.

PENGELLEY, E. T. *Circannual Clocks: Annual Biological Rhythms.* New York: Academic Press, Inc., 1974. 14 papers comprising the proceedings of a Satellite Symposium of the 140th Meeting of the American Association for the Advancement of Science in San Francisco in 1974. Includes a republication of a classic paper by William Rowan showing "daily increases in illumination . . . are conducive to developmental changes in the sexual organs" of Juncos. 523 pages.

PORTER, R., and G. M. COLLINS, eds. *Photoperiodic Regulation of Insect and Molluscan Hormones.* London Pitman, 1984. 18 research articles from the Symposium on Photoperiodic regulation of insect and molluscan hormones held at the Ciba Foundation, London, 1983. 298 pages.

REINBERG, A., and M. H. SMOLENSKY. *Biological Rhythms and Medicine: Cellular, Metabolic, Physiopathologic, and Pharmacologic Aspects.* New York: Springer-Verlag New York, Inc., 1983. Seven chapters by seven authors, intended as a textbook of modern, applied chronobiology. 305 pages, 148 illustrations.

RICHTER, C. P. *Biological Clocks in Medicine and Psychiatry.* Springfield, IL: Charles C. Thomas, Publisher, 1965. Originated from lectures on biological clocks in animals and in man. 109 pages.

ROSE, K. J. *The Body in Time.* New York: John Wiley & Sons, Inc., 1988. A science writer considers a potpourri of time courses for biological events including but not limited to sleep and circadian rhythms (sneezes, rates of development, development of an embryo, etc.) 237 pages.

SAUNDERS, D. S. *A Concise Introduction to Biological Rhythms.* Glasgow and London: Blackie, 1970. Written for "advanced (Scottish) undergraduates and those beginning a research career in biochronometry"; with a Glossary. 170 pages.

SAUNDERS, D. S. *Insect Clocks*. Oxford: Pergamon Press, 1976. Monograph on the fundamental properties of circadian rhythms and seasonal photoperiodism as they have been observed in insects; with a Glossary. 279 pages.

SCHEVING, L. E., F. HALBERG, J. E. PAULY, eds. *Chronobiology*. Tokyo: Igaku Shoin LTD., 1974. Proceedings of the first conference of the International Society for Chronobiology held in Little Rock.

SOLLBERGER, A. *Biological Rhythm Research*. Amsterdam: Elsevier Pub. Co., 1965. A book that "purports . . . to offer comprehensive information on biological rhythm research" and "a basic understanding of future problems." 461 pages.

STROGATZ, S. *The Mathematical Structure of the Human Sleep-Wake Cycle*. New York: Springer-Verlag New York, Inc., 1986. Using human raw data obtained from other investigators (and presented as a data bank of raster plots), the author analyzes some models and makes some simulations. He concludes that "all the models manage a qualitative fit to the data but on more rigorous tests they all fail." 230 pages.

STRUGHOLD, H. *Your Body Clock*. New York: Charles Scribner's Sons, 1971. Considers the "physiological clock in relation to rapid and drastic changes of the environment;" chapters consider sleep, shift work, air travel, and space missions. 94 pages.

SWEENEY, B. M. *Rhythmic Phenomena in Plants*. New York: Academic Press, 1969. Written for the "students of the future who will reveal the balance wheel of the biological clock." 147 pages.

WALKER, C. A., C. M. WINGET, and K. F. A. SOLIMAN, eds. *Chronopharmacology and Chronotherapeutics*. Tallahassee, FL: Florida A & M Foundation, 1981. Thirty-nine research articles from the "First symposium ever in the field of chronopharmacology;" International Symposium on Chronopharmacology and Chronotherapeutics, School of Pharmacy, Florida A & M University, Tallahassee, Florida. 417 pages.

WARD, R. R. *The Living Clocks*. New York: Mentor Books, 1971. An attempt to "tell the story of living clocks for readers who have no special background in science." 351 pages.

WEHR, T. A., and F. K. GOODWIN, eds. *Circadian Rhythms in Psychiatry*. V. 2 of Psychobiology and Psychopathology. Pacific Grove, CA: The Boxwood Press, 1983. Thirteen research articles. 270 pages.

WINFREE, A. T. *The Geometry of Biological Time*. New York: Springer-Verlag New York, Inc., 1980. Volume 8 of Biomathematics series, 290 illustrations. 530 pages.

WINFREE, A. T. *The Timing of Biological Clocks*. Scientific American Library, 1987. Two hundred thirty-three illustrations including color computer graphics. 199 pages.

INDEX